BUCKMINSTER FULLER

POET OF GEOMETRY

Written & Illustrated by Cole Gerst

BUCKMINSTER FULLER
POET OF GEOMETRY

Written & Illustrated by Cole Gerst

The works of Buckminster Fuller © The Estate of R. Buckminster Fuller.
The word DYMAXION is a trademark of The Estate of R. Buckminster Fuller;
used with permission. All rights reserved.
Climatron® is a registered Trademark of the Missouri Botanical Gardens.
The works of Isamu Noguchi copyright © The Isamu Noguchi Foundation.

Book design by Cole Gerst / option-g

Special Thanks: Greg Tomlinson, Lea Anne Clifton, Sue Gerst, John Ferry & The
Estate of Buckminster Fuller, Pablo Freund & The Buckminster Fuller Institute.

First Edition
Published by Option-G Visual Communication
option-g.com

IBSN 978-0-615-87344-2
Printed in China

BUCKMINSTER FULLER
POET OF GEOMETRY

Written & Illustrated by Cole Gerst

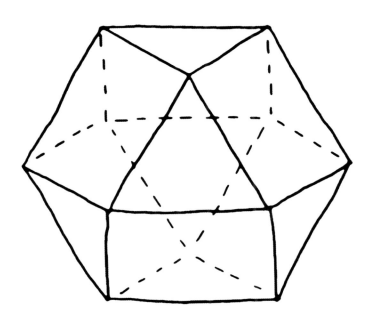

INTRODUCTION

Writing a book about one of the twentieth century's most brilliant minds isn't easy. Especially when you aren't a scientist, architect, or inventor, much less a writer. I didn't even set out to write a book, it just kind of happened. I always knew about Buckminster Fuller. I mean he did get an official stamp from the U.S. Postal Service! I was twelve years old when he died. I knew about him much like I think the majority of people know him, from his work with the geodesic dome. And although I had been inside of a handful of geodesic domes, Buckminster Fuller didn't always come to mind when I stood inside them.

My newfound fascination with "Bucky" happened organically. I had set about designing a series of tables based on one simple shape, the triangle. As a kid, I failed miserably in school at mathematics, but now I found myself studying geometry and loving it. It comes as no surprise to anyone who has studied Buckminster Fuller that his name came kept popping up in regards to the structural integrity of the triangle, or the icosahedron to be exact. Next thing you know, I'm reading every article, book and blog-post I can on him. I read every book *by* him I could get my hands on. There are a lot of Bucky related books out there.

So why did I decide there needed to be another book? Two reasons. First, I started by asking people how much they knew about Fuller. Their ages ranged from the teens well into the seventies. I figured some would know more than I initially did, or at least knew about the same. I was astounded that a small portion of people didn't even know who he was. Most knew about what I did before my intensive studies. My architect and artist friends had the most knowledge out of everyone. I found myself going on and on to anyone who would listen to me talk about all of his ideas and designs. Second, I saw a lack of books that might appeal to someone that might not know anything about him. The majority of books have no color or very little artwork. Due to the era in which Fuller lived, the majority of photos included are black and white. His ideas can be very complex, and his own writings are sometimes hard to follow. Once you dive into these books though,

they can be very rewarding. If I could just get more people to read about him, I would be happy, and his ideas would keep growing.

I decided to go about spreading the word by the best way I knew how, by drawing. I'm an illustrator and that is how I communicate. My goal is to get the next person inspired by Bucky and to go on to read more books by and about him. Then, hopefully, they will inspire someone else and so on. Whether it be his dedication to solving problems for humanity, his philosophy of "doing more with less", or simply inspiring creative thinking, I hope there is something you can take away from this book. I believe art and science go hand in hand and fostering creativity can lead to scientific breakthroughs. Buckminster Fuller epitomizes this belief.

"Now there is one outstandingly important fact regarding Spaceship Earth, and that is that no instruction book came with it."

R. Buckminster Fuller

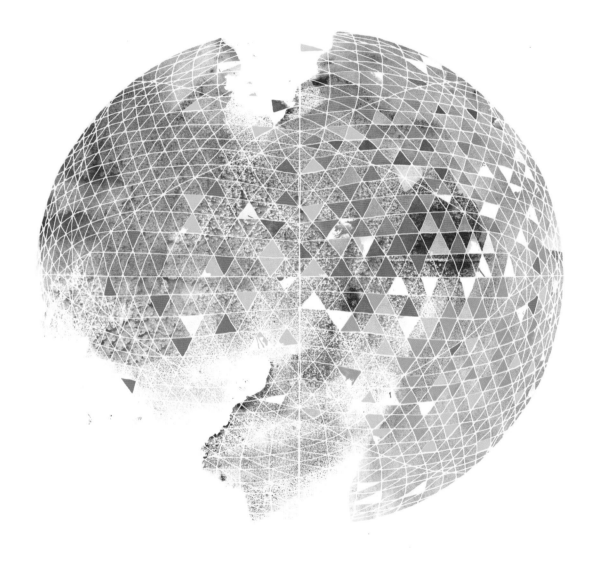

Spaceship Earth received a pilot on July 12th, 1895. On this day in Milton, Massachusetts, Richard Buckminster Fuller was born. Fuller's parents, Richard Fuller and Caroline Wolcott, were part of a family active in the politics of the time and known for being individualists, activists and public servants. "Bucky", as Buckminster came to be called, was also the grandnephew of Margaret Fuller a well-known American Transcendentalist. Margaret was a noted writer and critic and widely considered to have started the women's rights movement.

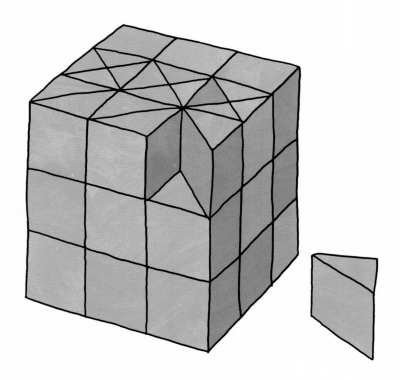

EARLY INFLUENCE

Bucky was born with very poor vision. Most everything he saw was only a blur of shapes and colors. Not until he was four years old could he see clearly. He was then fitted with glasses and saw what the world really looked like for the first time.

When Bucky turned six years old he attended the Froebel Kindergarten. In the early 1800's Friedrich Froebel created the concept of Kindergarten, which became the basis for modern education. He recognized that younger children have unique needs and learn quite differently than older people. Frobel harnessed the impulse of child's play and focused them into using that energy into a learning experience. Froebel created what he called "gifts" which are given to a child in succession for a unique learning experience. From each "gift" a different lesson can be achieved.

The "gifts" contained such simple objects as yarn balls, wood blocks, colored wooden tablets, triangular prisms, sticks, and rings.

It was with the lesson called "peas-works" where Bucky created a tetrahedral octet truss out of dried peas and toothpicks to the amazement of his teachers. This experience obviously made a mark on him as he went on to patent this creation in 1961 as the Octet Truss. It is interesting to note that other influential people such as Frank Lloyd Wright, Paul Klee, Piet Mondrian, and Helen Keller were also influenced or taught by the Froebel system.

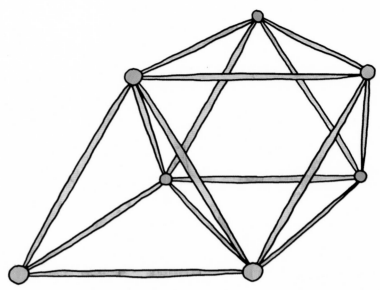

• Bucky's tetrahedral octet truss structure made at age six.

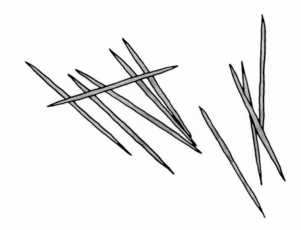

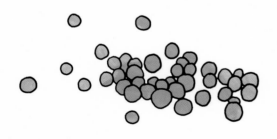

KINDER GARTEN TOYS
PEAS WORK.

13

• Lady Anne, a second-hand ship at Bear Island that Buckminster Fuller bought and refurbished.

BEAR ISLAND

Fuller went on to attend Milton Academy and later Harvard University (where he was expelled twice!), but it was when his Grandmother, Matilda Walcott Andrews, bought Bear Island around 1904 that we reach another important milestone in the early influence on Fuller's life.

Bear Island was located off the coast of Maine. Not a very big island, but big enough to roam for a day, Bucky soaked in the environment of the sparse, rocky island and the ocean. It was a refreshing contrast to living in a city and gave him a sense of freedom. Here he learned to navigate the waters on various types of boats and lived without running water and electricity. He would visit Bear Island as often as he could for the rest of his life.

· A young Buckminster Fuller rowing off the shore of Bear Island in his self-built canoe.

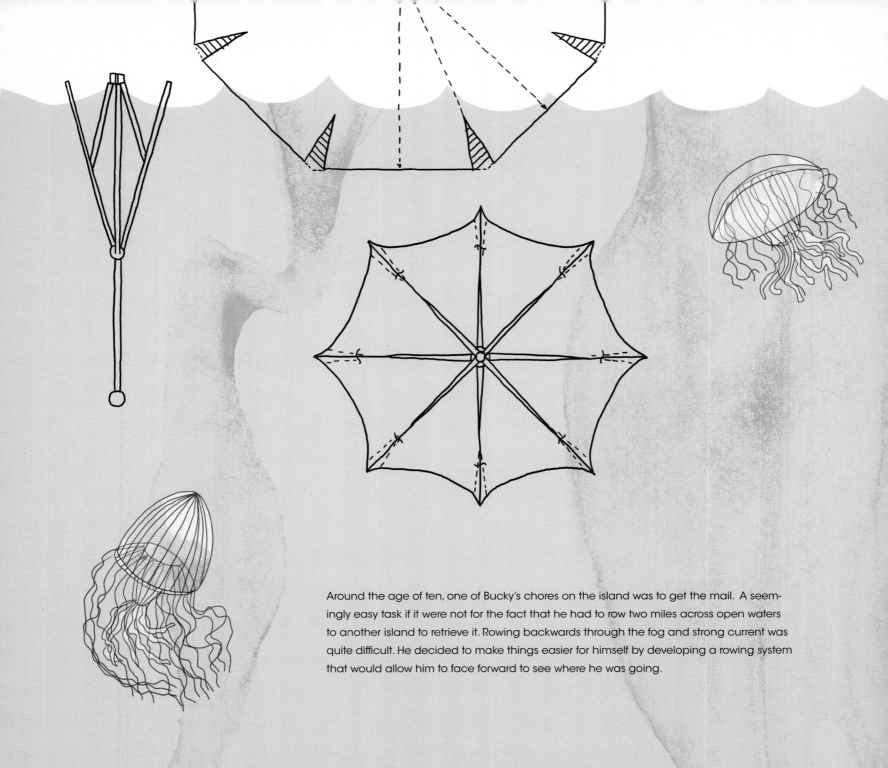

Around the age of ten, one of Bucky's chores on the island was to get the mail. A seemingly easy task if it were not for the fact that he had to row two miles across open waters to another island to retrieve it. Rowing backwards through the fog and strong current was quite difficult. He decided to make things easier for himself by developing a rowing system that would allow him to face forward to see where he was going.

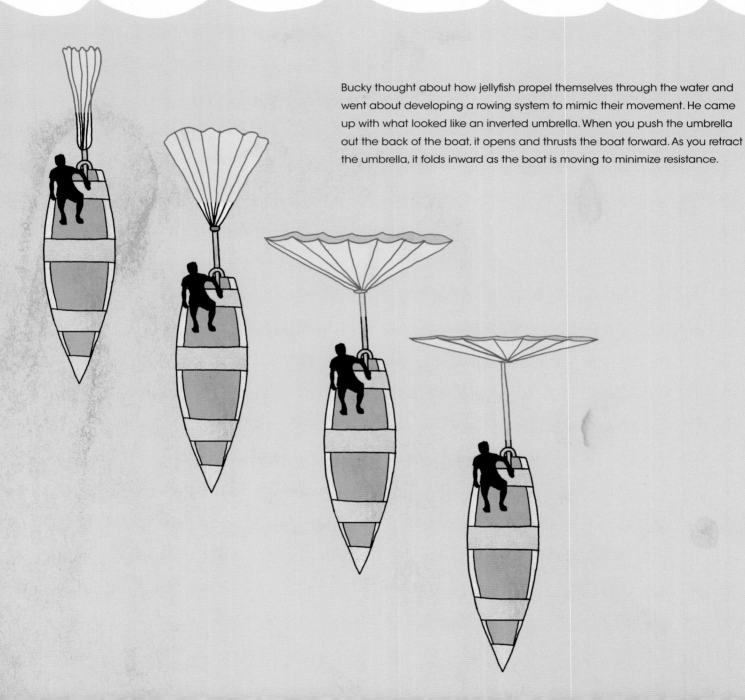

Bucky thought about how jellyfish propel themselves through the water and went about developing a rowing system to mimic their movement. He came up with what looked like an inverted umbrella. When you push the umbrella out the back of the boat, it opens and thrusts the boat forward. As you retract the umbrella, it folds inward as the boat is moving to minimize resistance.

THE NAVY

Sadness falls on the Fuller family as Bucky's grandmother passes away in 1906 and his father dies of complications from multiple strokes in 1908. This leaves Bucky with a lot more responsibility at home. Bucky finishes Milton Academy and attends Harvard, but the Harvard life is not for him. He works a few odd jobs until 1917.

The U.S. is in the midst of war when Fuller offers up a family boat to the Navy for use with him as Captain. He patrols off the coast of Maine in this tiny boat until the Navy retires her. Bucky loved his time on the sea. He ends up back at Harvard for Naval training. He excelled in the Navy and was very happy. Fuller learned much about shipbuilding and other technological advances during his time in the Navy. He believed that most developments in science could be directly linked to the desire for people to explore and reach new shores.

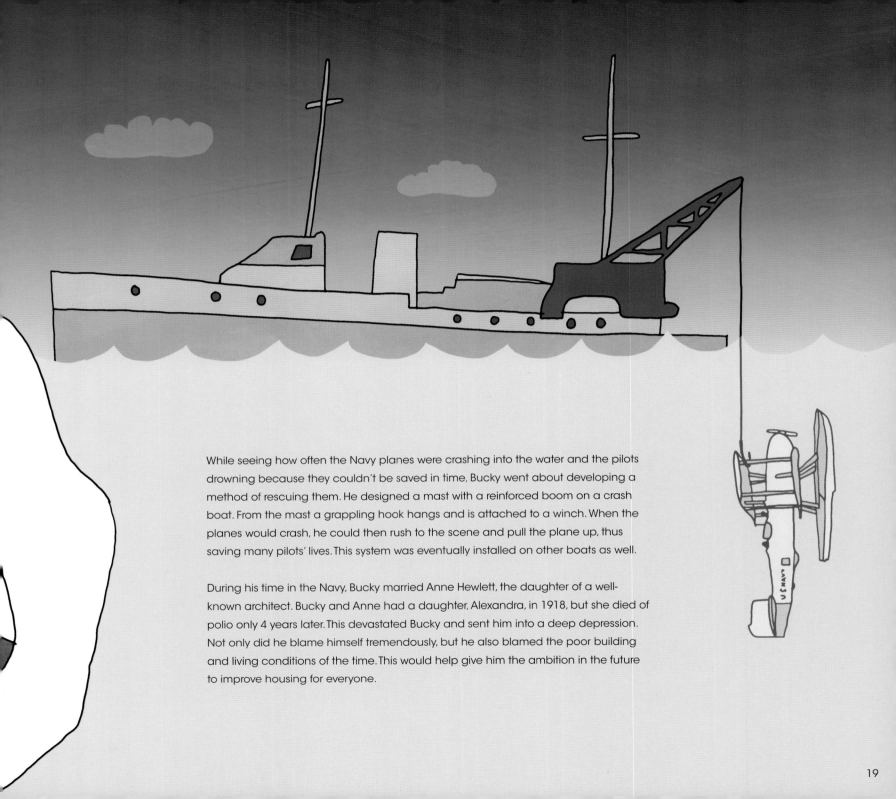

While seeing how often the Navy planes were crashing into the water and the pilots drowning because they couldn't be saved in time, Bucky went about developing a method of rescuing them. He designed a mast with a reinforced boom on a crash boat. From the mast a grappling hook hangs and is attached to a winch. When the planes would crash, he could then rush to the scene and pull the plane up, thus saving many pilots' lives. This system was eventually installed on other boats as well.

During his time in the Navy, Bucky married Anne Hewlett, the daughter of a well-known architect. Bucky and Anne had a daughter, Alexandra, in 1918, but she died of polio only 4 years later. This devastated Bucky and sent him into a deep depression. Not only did he blame himself tremendously, but he also blamed the poor building and living conditions of the time. This would help give him the ambition in the future to improve housing for everyone.

STOCKADE

Bucky's father-in-law had invented a new type of building system, but wasn't doing anything with it. Bucky thought it was quite good and wanted to help him develop it. The system was based on building blocks that were made of straw or other fibrous material that were mixed with cement and molded into a form. The form contained holes that could be filled with concrete, thus making a support pole within the structure. They were waterproof and could be sawed easily for special applications. They also came in various sizes and had different holes for different types of insertions, such as a horizontal beam.

Bucky was jobless at the time so they formed the Stockade Building Corporation in Brooklyn, New York. A patent was applied for and granted in 1927. They also patented a mold and process contraption for building the blocks. Their Stockade Building System soon went into production. Hundreds of buildings were made using this system.

Bucky and Anne soon moved to Chicago where, as President of the Stockade Building Corporation, Bucky oversaw production of structures in that area. The building system was advantageous because it could allow structures to be built faster and cheaper and with less labor, but with that came a lot of protest from the status quo building industry of the time. The labor force fought any attempt to modernize the process of building because it would ultimately, in their minds, cost them their jobs.

Thinking back on his daughter Alexandra and the poor living conditions they had, which he thought contributed to her death, he developed a deep contempt for the entire building industry. Eventually Bucky's father-in-law had to sell his share of the business to Celotex Corporation. Celotex also gained control of all the patents. Now Bucky was just an employee and very unhappy. He was fired from his job not long after Anne gave birth to Allegra, their second daughter.

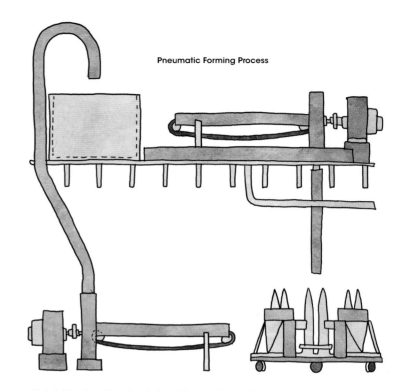

Pneumatic Forming Process

Material is cut, moistened and piped into a mold to create the blocks for housing structures.

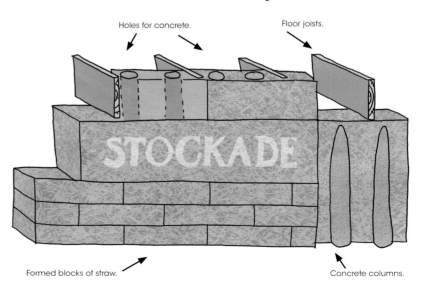

Stockade Building Structure

Holes for concrete.

Floor joists.

STOCKADE

Formed blocks of straw.

Concrete columns.

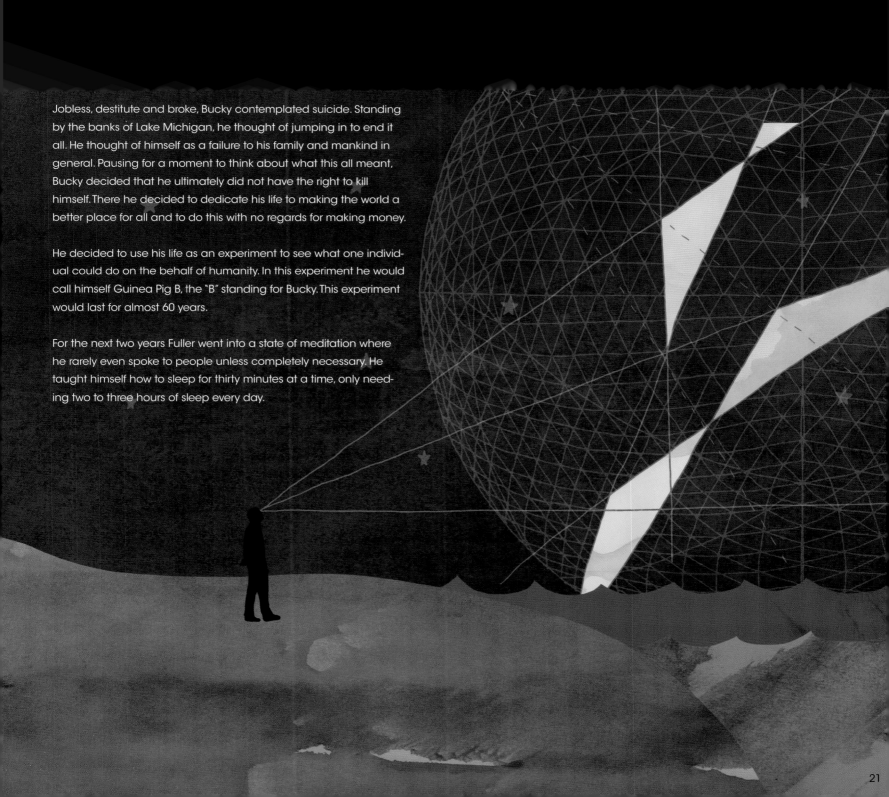

Jobless, destitute and broke, Bucky contemplated suicide. Standing by the banks of Lake Michigan, he thought of jumping in to end it all. He thought of himself as a failure to his family and mankind in general. Pausing for a moment to think about what this all meant, Bucky decided that he ultimately did not have the right to kill himself. There he decided to dedicate his life to making the world a better place for all and to do this with no regards for making money.

He decided to use his life as an experiment to see what one individual could do on the behalf of humanity. In this experiment he would call himself Guinea Pig B, the "B" standing for Bucky. This experiment would last for almost 60 years.

For the next two years Fuller went into a state of meditation where he rarely even spoke to people unless completely necessary. He taught himself how to sleep for thirty minutes at a time, only needing two to three hours of sleep every day.

DOING THE MOST WITH THE LEAST

Bucky concluded that out of all the advances that mankind was making, housing was falling far behind. He thought his own daughter had even died due to poor housing conditions. He devoted himself to making the living situations better for all. Using his past experiences with the Navy and The Stockade Building Company he started on his quest to do more with less.

Using less material equaled less money required to build. He looked at how sailing masts were built and how dirigibles used engineering and lightweight materials to achieve a lighter-than-air quality. Noticing how bicycle spokes used tension to create a rigid structure, he saw the benefits of applying this technique to building. The tetrahedron, he concluded, was the basic building block of the universe because of its strength of doing more with less.

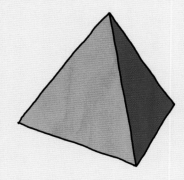

• A tetrahedron is a type of pyramid. In geometry it is defined as a polyhedron composed of four triangular faces, three of which meet at each vertex. It has six edges and four vertices.

• Bicycle spokes use tension to create a rigid structure.

• With a ship's mast, elements are suspended from above instead of being rested on support from below.

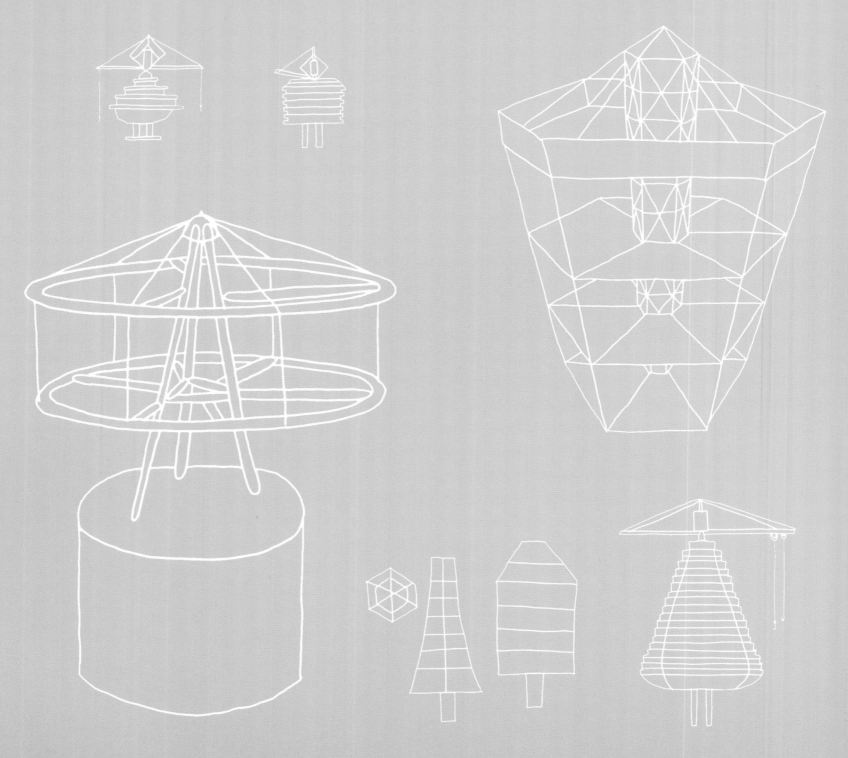

LIGHTFUL HOUSES

Using all of his new research and philosophy, Bucky went on to exploring new ways to apply these applications to housing. He began by designing from the inside out, basically engineering the support structure for what he soon called "Lightful Houses". He later named these 4D houses, after the fourth dimension, time.

He experimented with a chassis design similar to what he had seen in airship landing pylons. From there the rest of the structure would essentially be hung so no walls actually were load bearing. During this time, around 1928, Bucky submitted a patent application for a 4D house. He later withdrew it and continued working on the design.

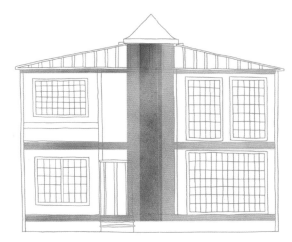

• Early 4D House used a rectangular floor plan, but the patent application was pulled by Fuller to further develop his idea.

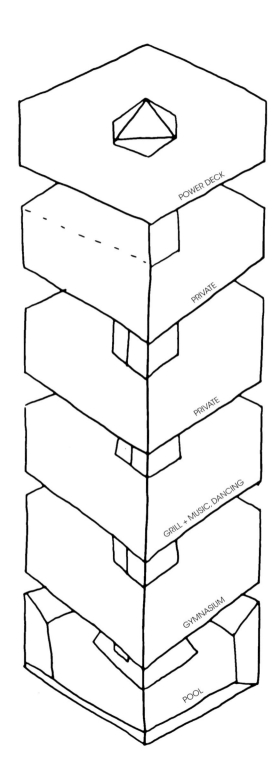

POWER DECK

PRIVATE

PRIVATE

GRILL + MUSIC, DANCING

GYMNASIUM

POOL

Using the hexagon, which is comprised of six triangles, Bucky continued developing floor plans and building designs. He decided the most efficient way to build is to build up, stacking rooms on top of each other. These were called "Lightful Towers", and although they might resemble more modern apartment complexes, they are actually made for a single family.

These towers were lightweight and could be erected in one day. They would be completely independent of the power grid and sewage system. They would also generate their own light and heat and even came complete with built in furniture and a pool. Each floor had its own purpose such as for the power system, storage or living. The bottom floor was the pool, and an elevator ran to each floor through the core of the building. On top would be a radio tower for communication and even a way to harness the wind for energy.

He believed that someday soon society would depend on renewable sources of energy. He noted that the earth only had so much land as most of it is water, and the population would soon explode. Cheap and efficient housing should be available for everyone.

• A typical hexagonal floor plan consisting of an elevator shaft in the middle. Each floor had its own purpose such as a swimming pool on the bottom, a gymnasium, private areas and a deck on the very top

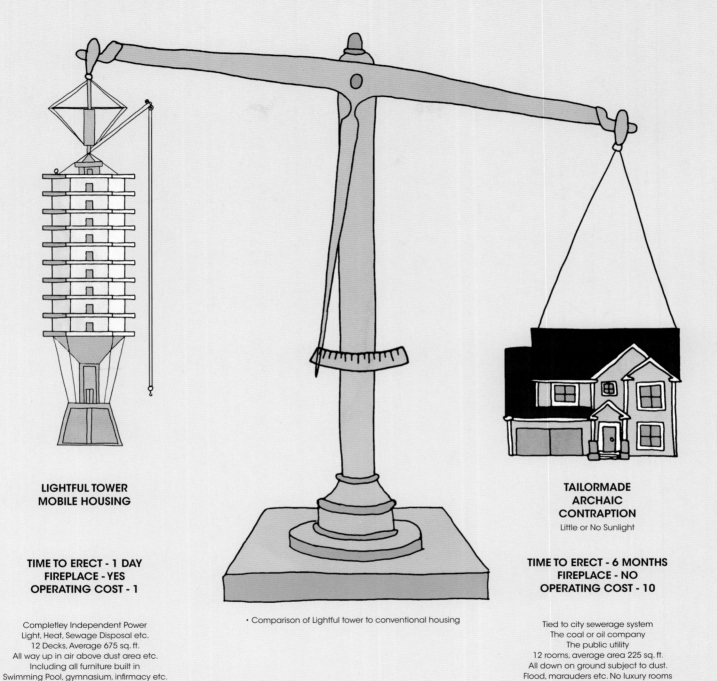

**LIGHTFUL TOWER
MOBILE HOUSING**

**TIME TO ERECT - 1 DAY
FIREPLACE - YES
OPERATING COST - 1**

Completley Independent Power
Light, Heat, Sewage Disposal etc.
12 Decks, Average 675 sq. ft.
All way up in air above dust area etc.
Including all furniture built in
Swimming Pool, gymnasium, infirmacy etc.
Free of Land

· Comparison of Lightful tower to conventional housing

**TAILORMADE
ARCHAIC
CONTRAPTION**

Little or No Sunlight

**TIME TO ERECT - 6 MONTHS
FIREPLACE - NO
OPERATING COST - 10**

Tied to city sewerage system
The coal or oil company
The public utility
12 rooms, average area 225 sq. ft.
All down on ground subject to dust.
Flood, marauders etc. No luxury rooms
No furniture included.
Tied down to Land

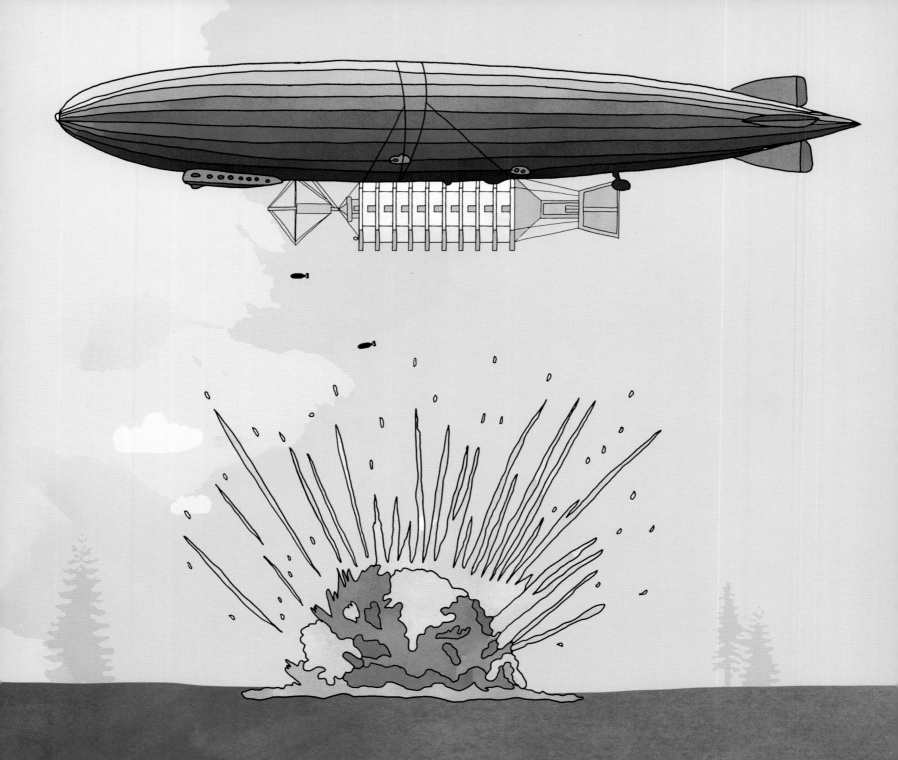

One of his most radical proposals was that these houses could be delivered anywhere in the world by Zeppelins. They would be mass-produced in a factory similar to what Henry Ford was doing with the car at the time. Since they could be self-sustaining they could even be put on the North Pole! He proposed that these Zeppelins carry the building to a site, drop a bomb that would explode and create the hole in which the building was "planted". Men on the ground would shore up the building while the hole around it was being filled with concrete.

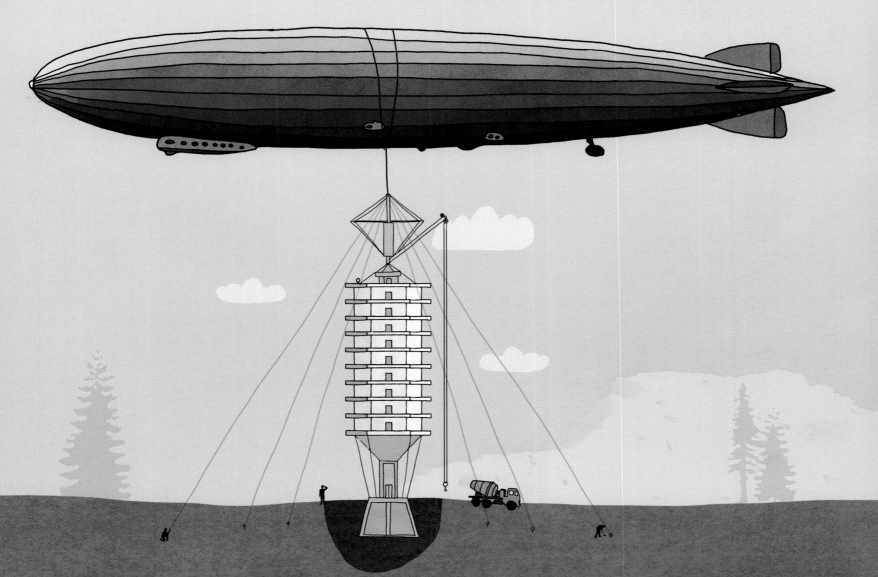

LIGHTFUL HOUSES
VISION OF A UNITED WORLD

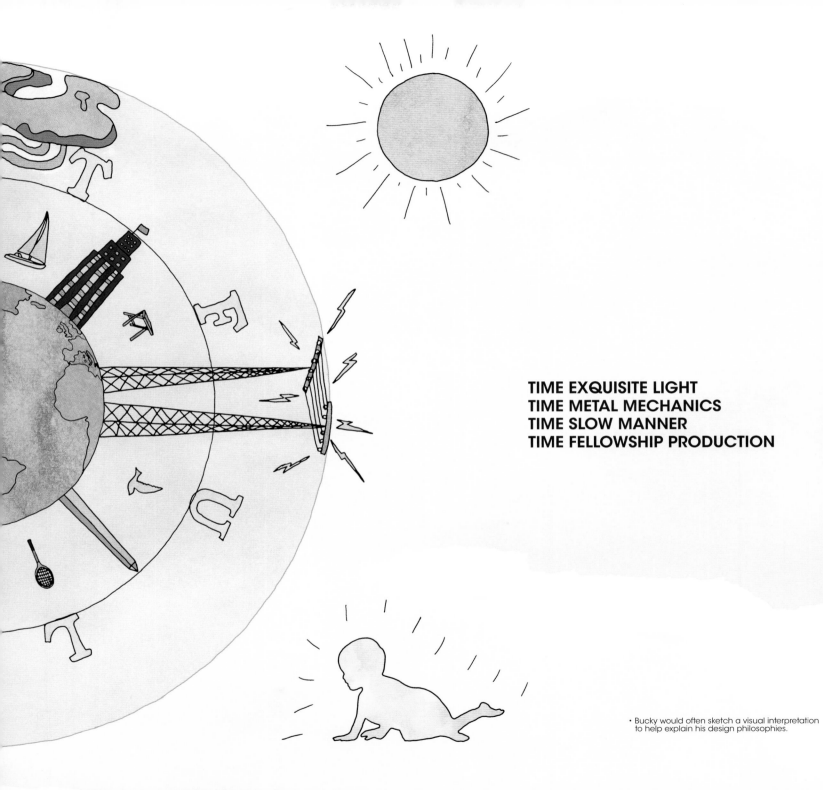

TIME EXQUISITE LIGHT
TIME METAL MECHANICS
TIME SLOW MANNER
TIME FELLOWSHIP PRODUCTION

• Bucky would often sketch a visual interpretation
  to help explain his design philosophies.

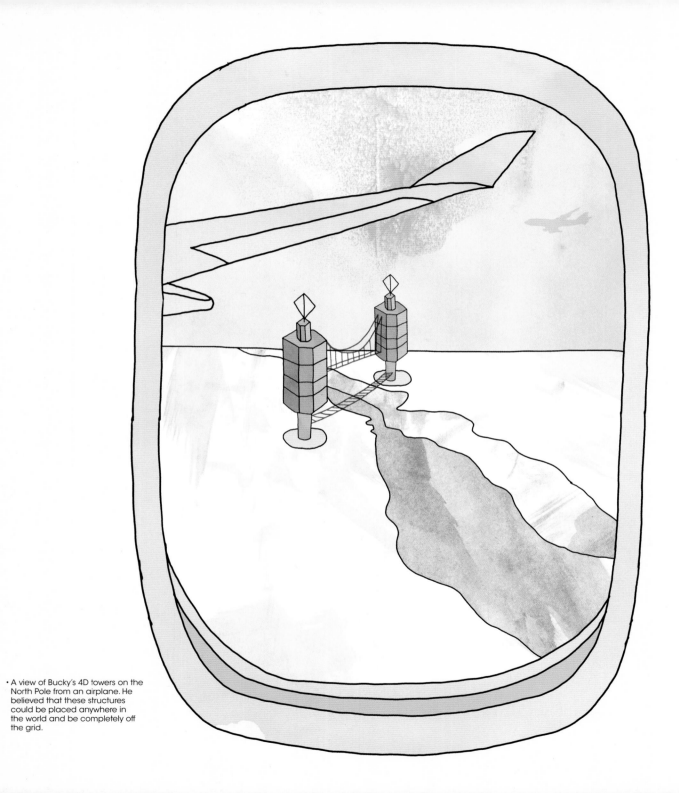

• A view of Bucky's 4D towers on the North Pole from an airplane. He believed that these structures could be placed anywhere in the world and be completely off the grid.

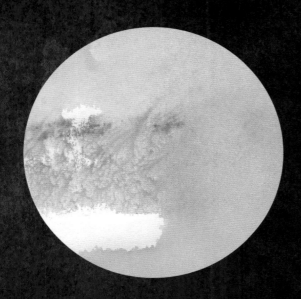

Many of these ideas were not even possible in the 1920's, and Bucky knew it. Some of the materials needed to achieve these concepts hadn't even been invented yet. The lightweight metal he needed wouldn't be available for another 25 years according to his calculations. He was designing it in the hopes that technology would catch up with his ideas if he could manage to convince everyone that a new way of living was needed.

Some of Bucky's radical thinking and plans for the towers, now called 4D towers, were published in an article in the Chicago Evening Post in 1928. This article started to give Bucky some recognition as he continued to streamline his designs.

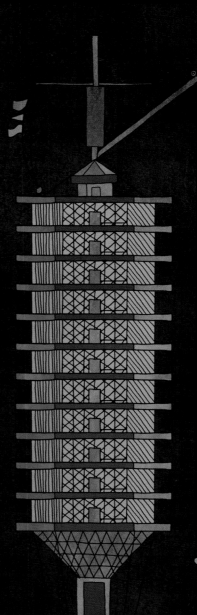

4D-TIME LOCK

After the recognition from the Chicago Tribune article, Bucky began working on a single story version of the tower. The floorplan was still hexagonal in shape, and this was divided into rooms that were triangular in shape. There was still a central mast where the elevator was located, and the bathrooms were located toward the center as well.

In the ground would be fuel, water and all the power generating equipment, which would be delivered up the central mast. On top of these rooms was a deck that was covered for protection from the elements. The cover also deflected wind. The walls still hung from the mast as in the 4D towers. The floors consisted of a netting that was sandwiched between a material to form the floor. Of course all the bathrooms and modern appliances were built in, as well as furniture. This house was meant to be mass-produced at a significant savings from creating a similar size house of the day.

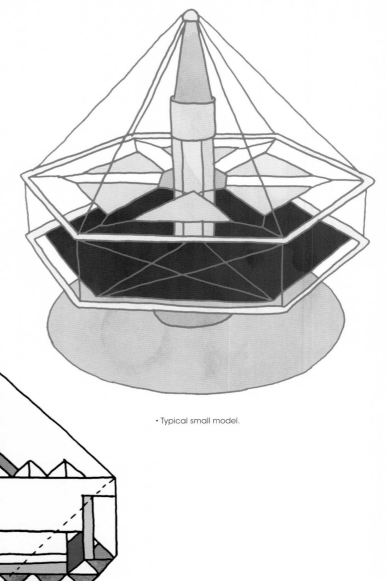

• Typical small model.

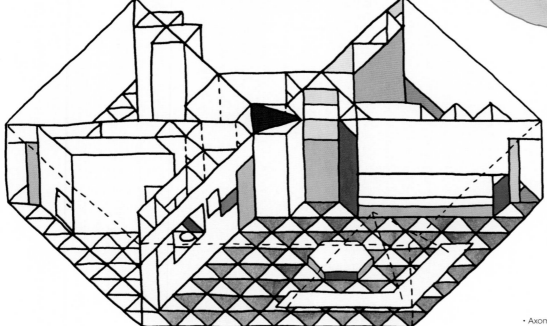

• Axometric plan showing room layout.

34

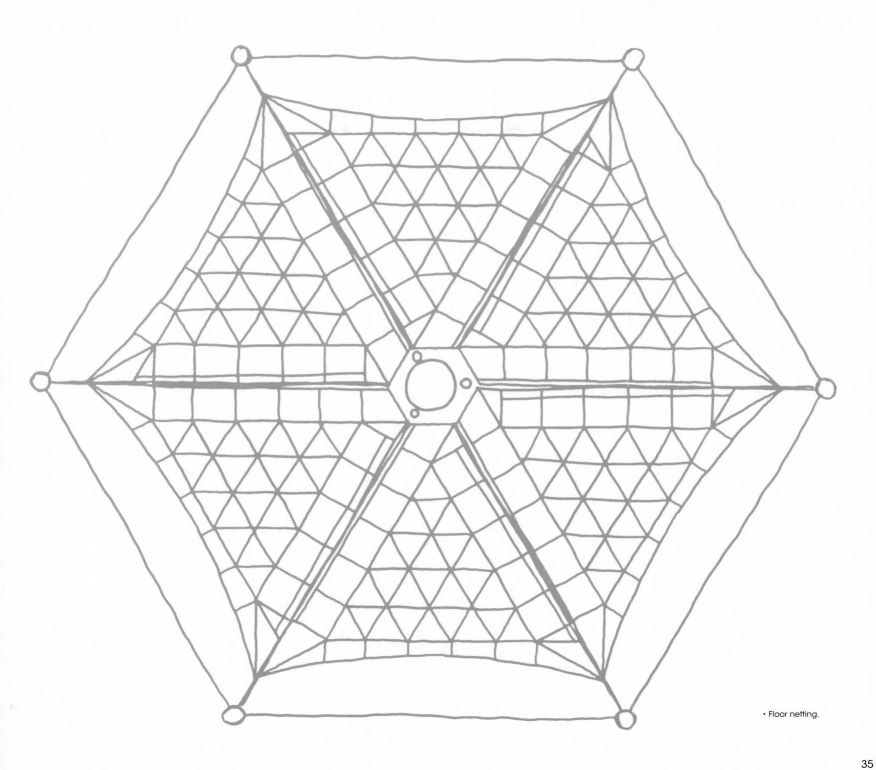

• Floor netting.

35

Bucky recieved some investment in his housing project and with this money started to make a model of the 4D house. First he secured patents and began writing all of his thoughts down. He wrote about using mass production for creating shelter and how the building industry had failed to take advantage of new technology.

He entitled his writings as "4D Time Lock". His paper was printed and distributed to architects, including the American Institute of Architects. He also had copies distributed to other prominent people such as Henry Ford. The AIA dismissed his designs because of their mass production approach. Of course they would be against this approach, it made architects almost obsolete in the design and building process.

• One of many 4D logo ideas.

"IN WHICH THE GREAT COMBINATION IS REVEALED, IF THOUGHTFULLY FOLLOWED IN THE ORDER SET DOWN, AWAITING THE CLICK AT EACH TURN."

• Text from Bucky's "4D Time Lock" book cover.

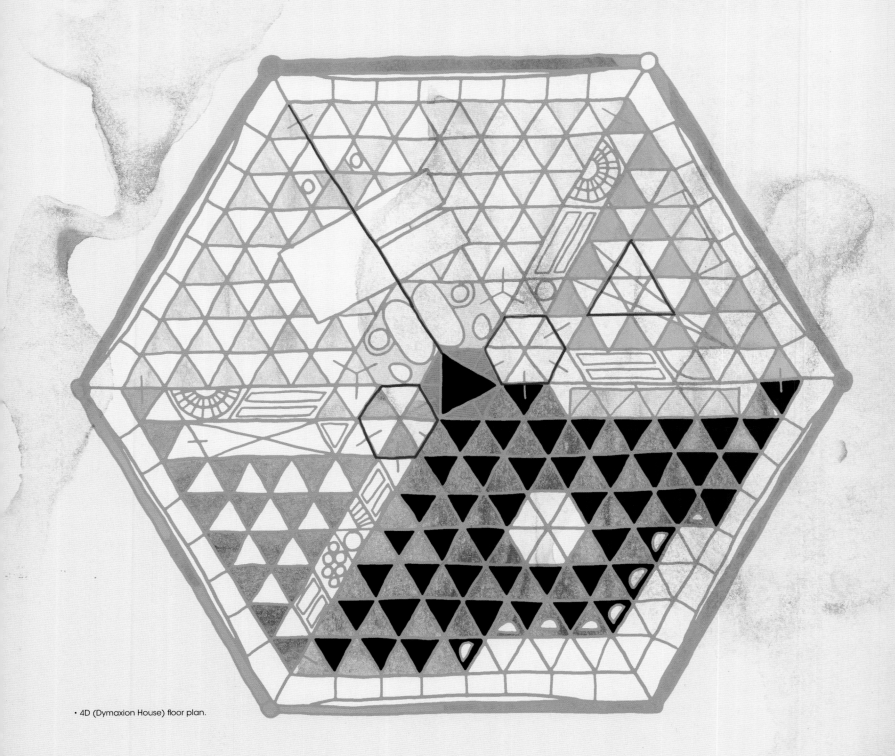

• 4D (Dymaxion House) floor plan.

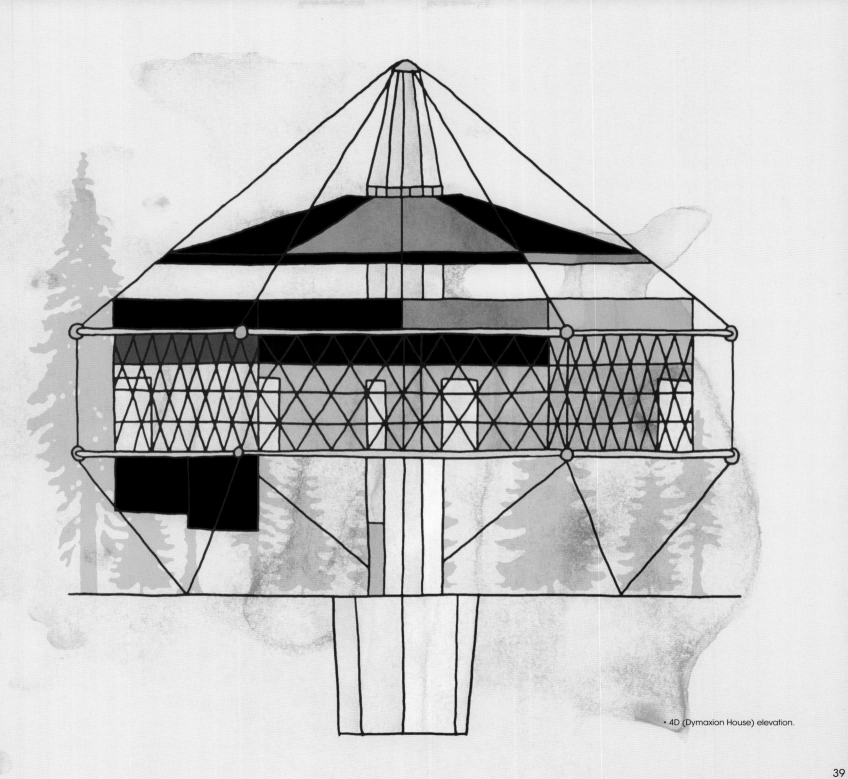

• 4D (Dymaxion House) elevation.

• A Dymaxion logo design.

DYMAXION

Bucky was still barely speaking to anyone at all. He had mainly been corresponding with people via letter and his manuscript. With his finished model, he began speaking again by having lectures at various clubs about the 4D house.

In 1929, the department store Marshall Fields got wind of Bucky's house design. It so happened that they had a line of modern furniture for sale and thought that Bucky showing his new modern housing system might help spark sales. Marshall Fields did not, however, like the name of the 4D house.

An advertising representative by the name of Waldo Warren met with Bucky to talk to him about changing the name. Mr. Warren was well known and had coined the term "radio". Waldo combined three words that he thought were representative of Bucky and his invention, as well as being more dramatic. These words were dynamic, maximum and ion. From here "Dymaxion" was born, and later became Bucky's trademark most people today are familiar with.

• Fuller speaking at Marshall Fields Department Store.

41

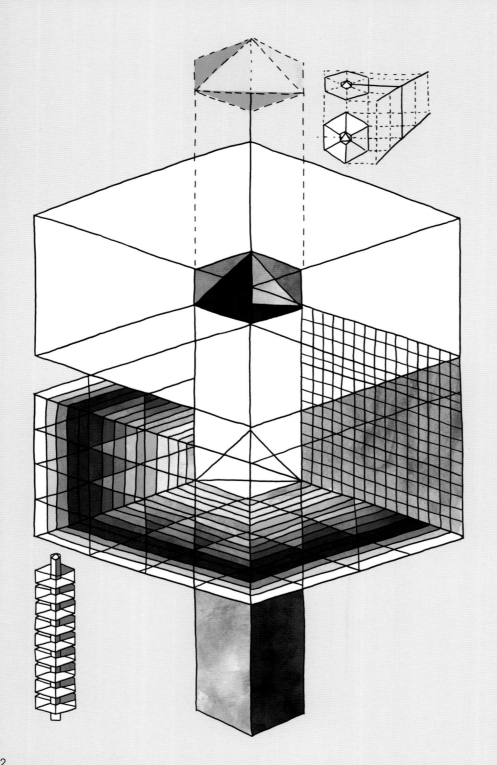

4D
TIME INTERVAL 1 METER

CHART INDICATING 4D COLOR PROGRESSION
FROM DARKNESS THROUGH YELLOW OF
DAWN (1ST LIGHT) THROUGH NATURAL GREEN
AND MECHANICAL RED TO THE BLUE OF THE
UNIVERSE AND THE COMPLETE LIGHT OF ETERNITY

· Measuring time from the central axis of the 4D tower.

• Interior rendering of the 4D house.

DYMAXION MOBILE DORMITORY

In the early 1930's, Bucky designed a version of the Dymaxion House as a shelter for the migrant farmers in Russia. The country was having an industrial revolution at the time and in need of cheap and easy to build structures. Bucky had visited the Ural Mountains in Russia and had seen the poor, rugged conditions and the lack of materials, such as metal.

He decided he would design something that could utilize the main local material, which was wood. The floor was rope and packed with grass, and then a compressed sawdust material would finish out the surface of the floor. The walls were fabric panels and were suspended by a central mast. It still maintained his hexagonal floor plan. At the peak of the roof was a ventilation system to control air circulation. It even was designed to use their tractors as a source of power for heating and light.

• Structural studies of the Dymaxion Mobile Dormitory.

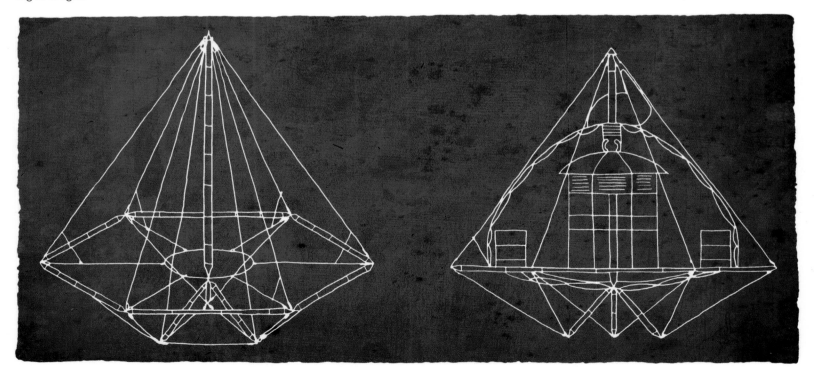

FULLER 4D TOWER GARAGE

THIS IS A SAMPLE OF MECHANICAL PERFECTION THAT
OLD CITIES WOULD NEVER BE ABLE TO PERMIT
THAT IS WHY 4D STARTS WITH COUNTRY WORK

SELF PARKING
KEEP ON UP TILL YOU FIND A SPACE
UP + DOWN
RAMPS NEVER CROSS

PLAN SHOWING TOWER SUPPORT + ELEVATOR
TWO CIRCLES OF PARKING

SEPARATE RAMPS UP AND DOWN

CENTRAL TOWER SUPPORTS AND HOUSES ELEVATOR
TO AND FROM CARS

FENCE AROUND BOTTOM TOLL HOUSE
FLOORS SUPPORTED BY CABLES

COULD BE MADE 100 DECKS HIGH AND BE
COLOSSALLY BEAUTIFUL

• Text taken from a circa 1928 Buckminster Fuller sketch of proposed garage.

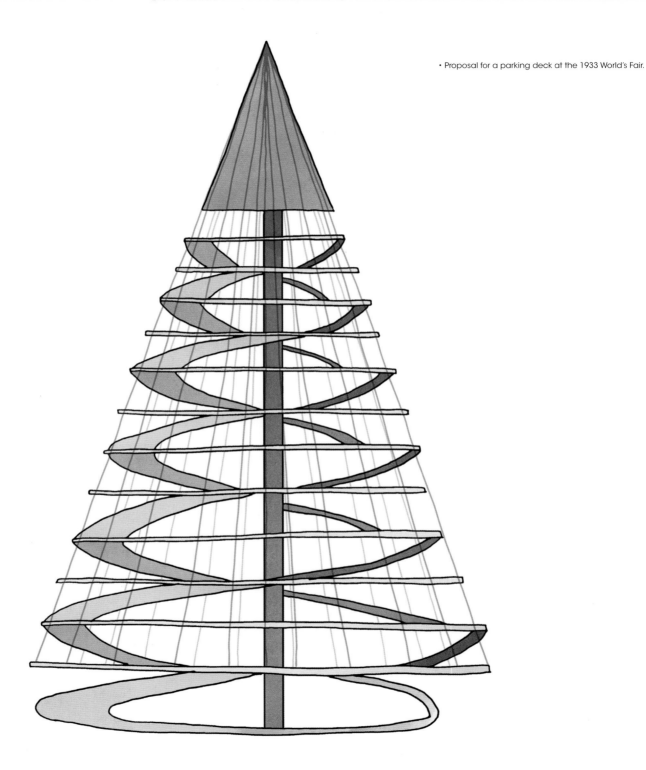

• Proposal for a parking deck at the 1933 World's Fair.

• Isamu Noguchi

48

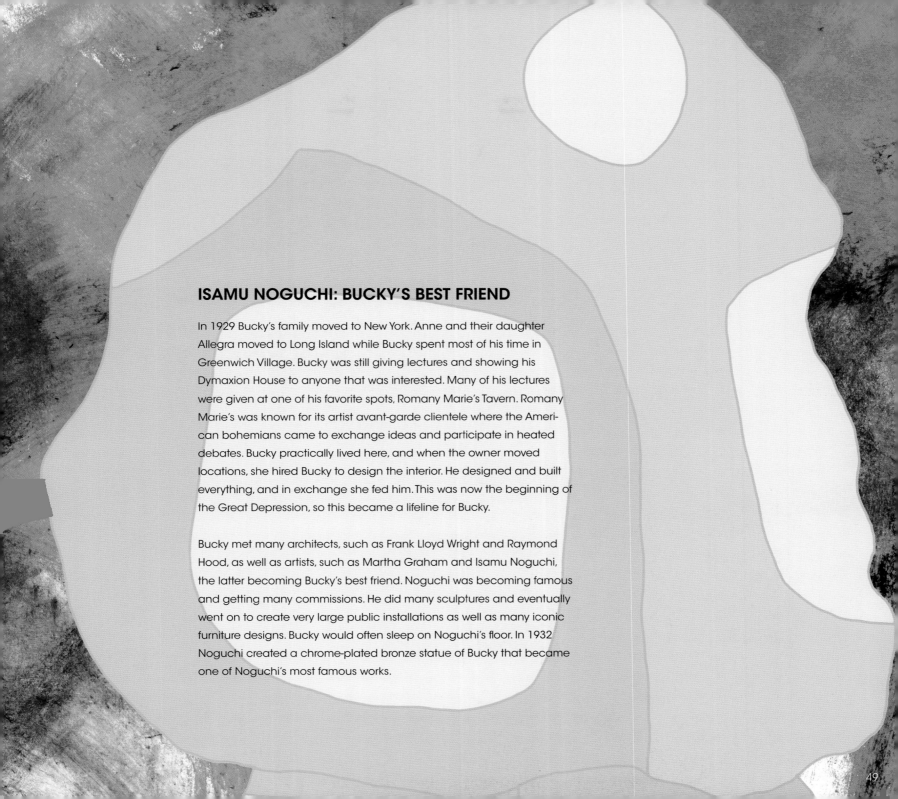

ISAMU NOGUCHI: BUCKY'S BEST FRIEND

In 1929 Bucky's family moved to New York. Anne and their daughter Allegra moved to Long Island while Bucky spent most of his time in Greenwich Village. Bucky was still giving lectures and showing his Dymaxion House to anyone that was interested. Many of his lectures were given at one of his favorite spots, Romany Marie's Tavern. Romany Marie's was known for its artist avant-garde clientele where the American bohemians came to exchange ideas and participate in heated debates. Bucky practically lived here, and when the owner moved locations, she hired Bucky to design the interior. He designed and built everything, and in exchange she fed him. This was now the beginning of the Great Depression, so this became a lifeline for Bucky.

Bucky met many architects, such as Frank Lloyd Wright and Raymond Hood, as well as artists, such as Martha Graham and Isamu Noguchi, the latter becoming Bucky's best friend. Noguchi was becoming famous and getting many commissions. He did many sculptures and eventually went on to create very large public installations as well as many iconic furniture designs. Bucky would often sleep on Noguchi's floor. In 1932 Noguchi created a chrome-plated bronze statue of Bucky that became one of Noguchi's most famous works.

• Noguchi's chrome plated statue of Bucky.

Noguchi and Fuller would also collaborate from time to time. In 1930 Bucky bought the magazine *T-Square* by cashing in his life insurance policy. He changed the magazine's name to *Shelter*. Noguchi would contribute to the magazine by creating drawings of Bucky's ideas or having sculptures appear on the cover. The last issue of *Shelter* featured Noguchi's "Miss Expanding Universe".

In some of the issues we can see some of Bucky's studies on transportation and streamlining. A new focus for Bucky was a vehicle he called the "4D Transport Unit". Noguchi made some 3D models of these units out of gypsum. *Shelter* went out of business in 1932 but we can already see where Bucky's next big project was heading.

• One of Noguchi's sculptures, "Miss Expanding Universe", on the cover of *Shelter Magazine*.

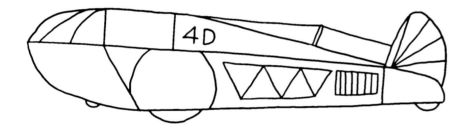

• Noguchi would often collaborate with Bucky by making gypsum models based off of Bucky's concepts. Bucky would go on to publish these models in his magazine, *Shelter*.

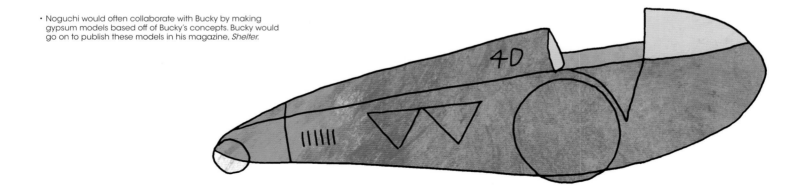

• Noguchi bridged the gap between art and furniture with pieces such as this 1948 coffee table simply called the "Noguchi Table".

• Designed by Noguchi beginning in 1951 and handmade for a half century by the original manufacturer in Gifu, Japan, the paper lanterns are a harmonious blend of Japanese handcraft and modernist form.

• Many major metropolitan cities commissioned Noguchi for public sculptures, including the 24 foot tall "Red Cube" in downtown Manhattan.

ON THE MOVE

Building a vehicle wasn't a new idea for Bucky. He had been dreaming of an aerial vehicle for many years. Instead of using rotating propellers as the airplanes of the day did, Bucky had been theorizing about jet engines that hadn't even been invented yet. He had designed a house that could be transported anywhere in the world, but how were those people to get around if there were no roads too get them too and from their home?

Bucky's first design was of a vehicle that could not only be driven like a car, but it could also float and travel by water. Oh, and if you just inflated the stored wings, you could just simply fly wherever you were going! It had two front wheels and only one back wheel. It would steer from the rear wheel like a fish swims. These designs were created before the coining of the phrase "Dymaxion" and were then simply called 4D Transport.

After closing his magazine *Shelter*, Bucky decided to pursue actually building a prototype of his vehicle. He knew that at the time, his air, land and sea vehicle wasn't practical and went about building a new kind of automobile. Instead of going out and trying to get funding, building models and lecturing, Bucky went straight to work in building a full scale working model. He already had a little help from a few loyal believers and then hired well-known yacht and seaplane builder, Star-ling Burgess. The day Franklin D. Roosevelt declared the country broke from the Depression, Bucky started building his Dymaxion Car.

• 4D Transportation Unit with inflatable wings.

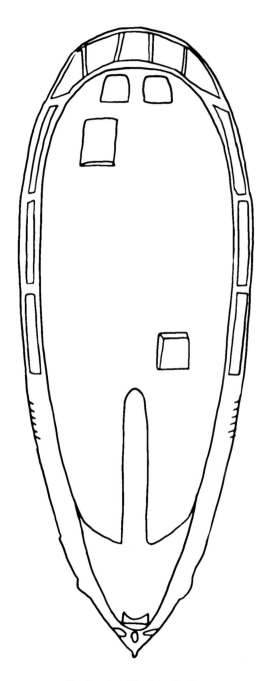

A Rolls Royce factory that had recently opened in the United States was forced to close due to the Depression. Bucky hired mechanics from this factory to work on his newest enterprise. The new design still had 3 wheels and the frame was made of wood and clad in aluminum. It had front wheel drive and was steered by the rear wheel, as a boat is steered by a rear rudder. The car could turn 360 degrees on itself, essentially circling in place. Henry Ford gave Bucky a V-8 engine at a huge discount that was installed. The Dymaxion Car could go up to 120 mph, which was much faster than the Ford models of the time because it was much lighter and more aerodynamic. It was therefore much more fuel-efficient as well. The car could hold up to eleven passengers.

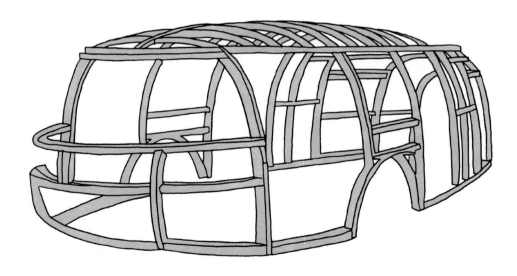

• Top view plan of the Dymaxion Car.

• The frame of the car was made out of wood, then clad in aluminium.

• Bucky with his first complete Dymaxion Car. Note the rearview mirror on the top. The driver would look up out of a skylight to view it.

Bucky and his crew would go on to build three separate prototypes of the Dymaxion Car, improving it every time. The car gained much attention, as it was a radical departure from the car of the day. Many claimed the car would help bring the United States out of the Depression. Bucky even got an invite from the Automobile Show in the Madison Square Garden, but Chrysler forced him out in a sly political move. That didn't stop Bucky from driving the car around outside of the car show to the delight of the crowd.

In 1933 an English Colonel came to the United States representing a potential buyer for the car. He was being driven in the Dymaxion Car by a famous racecar driver of the time near the entrance to the Chicago World's Fair when they were involved in a car accident. Apparently, a car containing a local politician had challenged them to a race. The Dymaxion Car was sideswiped causing it to flip. The driver was killed and the Colonel injured. Before the actual cause of the accident was reported in the papers it was too late, the Dymaxion had already been labeled as dangerous. There was a cover-up and the politician was never acknowledged as being the cause of the wreck. Another Dymaxion Car was eventually finished and shown at the Chicago Fair, but investors shied away due to bad publicity. That would be the last Dymaxion Car to be built even though Amelia Earhart, among others, had ordered one.

Both in 1943 and 1950 Bucky presented designs for a smaller version of the Dymaxion car at the request of Henry Kaiser, the well-known American industrialist. Both versions could seat four across. The cars were still three-wheeled and had the basic aerodynamic look of the original. Kaiser decided that they were too expensive and radical from a design perspective and never advanced beyond the drawings.

• A smaller version of the Dymaxion Car proposed in 1950.

• The second Dymaxion Car next to a Ford model from the same year. Both cars used the same engine. Note the rearview mirror was replaced with a periscope.

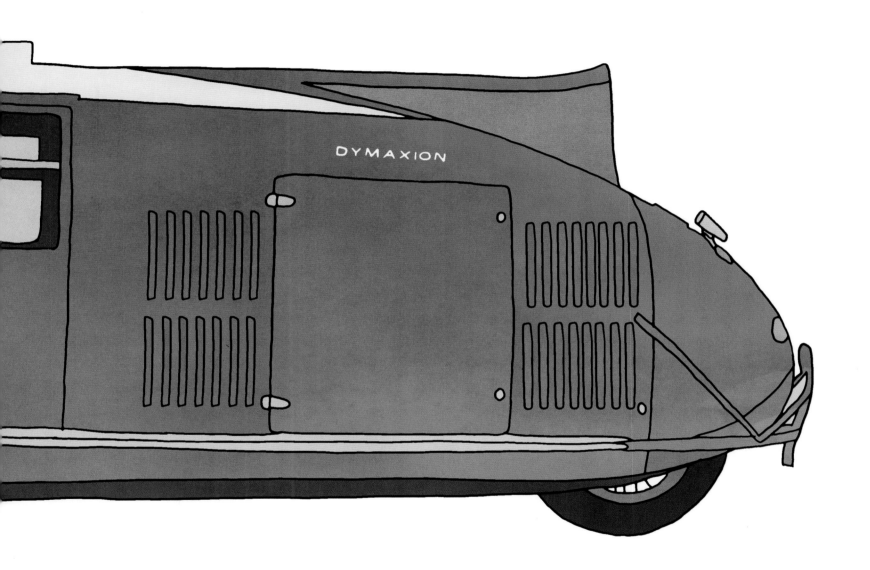

DYMAXION

• The third and last version of the Dymaxion Car.

INTEGRATION

In 1937 Bucky was already onto producing a new industrialized bath-room that could be mass-produced. It came in four main sections, but contained everything you needed in a typical bathroom. Built in were a sink, bathtub, toilet, lighting, ventilation, and plumbing. The material was die stamped metal that would be pieced together on site. These Dymaxion bathrooms would of course go into the Dymaxion House, but would also be used to replace older bathrooms in other houses and apartments.

The Phelps Dodge Corporation was to put the Dymaxion bathrooms, also known as the Integrated Bath, into production, but there was a lot of resistance from plumbers from the fear that they would lose their jobs. World War II would also put the production on hold and it never took off.

Bucky also had a few other ideas that could be incorporated into bathrooms. One of these ideas was a "fog gun". Noting the enormous waste of water in washing dishes or bathing, these fog guns would use compressed air with a small amount of water and no soap to quickly wash us clean. This would save water and eliminate a lot of pollution. Many of these ideas never caught on, but you can see some of Bucky's influence on modern bathrooms in airplanes and trains.

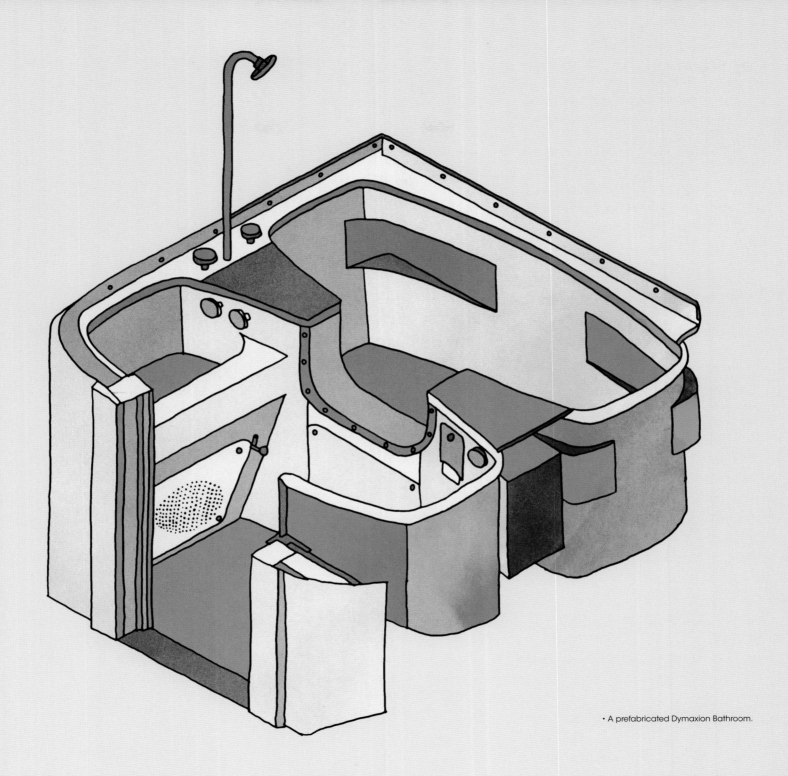

• A prefabricated Dymaxion Bathroom.

AN ADVENTURE STORY OF THOUGHT

Bucky had been writing. He was putting down his theories and vision into essays on future prosperity. He coined the term "ephemeralization" to refer to the ability of technological advancement to do "more with less". He referred to Henry Ford's assembly line as an example of how this process can lead to better products and productivity. He put all of these observations into a book called <u>Nine Chains to the Moon</u>. The title referred to the fact that if the entire population, at the time, stood on each others shoulders, they could reach back and forth to the moon nine times.

A lot of the book was devoted to Albert Einstein. Einstein had published an article in the New York Times in 1930 that deeply moved Bucky. In one chapter Bucky explains the formula $E=mc^2$ in layman's terms calling it "Mrs. Murphy's Horsepower". Bucky had first explained his interpretation of the mass-energy equivalence in a telegram to his friend Isamu Noguchi.

The book was all ready to be published, but when reading the chapter on Einstein, the editor declined to print it. Einstein had mentioned in an article after publishing $E=mc^2$, that there were only ten people in the world that might be capable of understanding his ideas. Bucky was not on that list, and the publishing company was afraid of being guilty of printing fraudulent material.

This was an obvious blow to Bucky and a huge disappointment. Luckily, Einstein had received a copy of Bucky's writings about him and wanted to meet him. Not only did Einstein approve of the writings, but Einstein told Bucky he didn't think anyone would have any practical applications for his formula. Bucky appeared to have found them. <u>Nine Chains to the Moon</u> was published in 1938. Many influential people read Bucky's book, and wrote him about what they thought of it. One of these people was Frank Lloyd Wright. Wright called him "extraordinarily sensitive".

• Bucky corresponded with and met both Albert Einstein & Frank Lloyd Wright.

DYMAXION DEPLOYMENT UNIT

Farmers use grain bins to keep grain safe from weather, bugs, and rodents. These corrugated metal structures are inexpensive and fireproof as well. Bucky came across these bins on a trip through the Midwest, and thought they would make a good foundation for cheap housing. His Dymaxion House hadn't gained much traction, but he was still looking for a feasible solution to make inexpensive housing for the masses. He approached the Butler Manufacturing Company about modifying their bins for housing. They liked the idea and soon the structures were being made.

The units used the galvanized metal to its full advantage by bending the roof into a curve while still maintaining strength. The units were easy to ship and build. They were to be built from the top down, so the roof was hoisted in the air and the walls were then filled in below. Bucky used this method in his future building projects. The units were insulated and furnished. Corrugated walls or curtains could create rooms within the unit. Multiple units could be added simply by cutting a door in the walls and attaching.

This was the early forties and the United States was in the middle of World War II. While there was still resistance by the building industry for prefabricated, low cost housing, the military was in need of shelters that were easy to build and easily transported overseas. Hundreds of Bucky's units were bought by the military and put in use as the Dymaxion Deployment Unit (DDU) until a metal shortage put a halt to production.

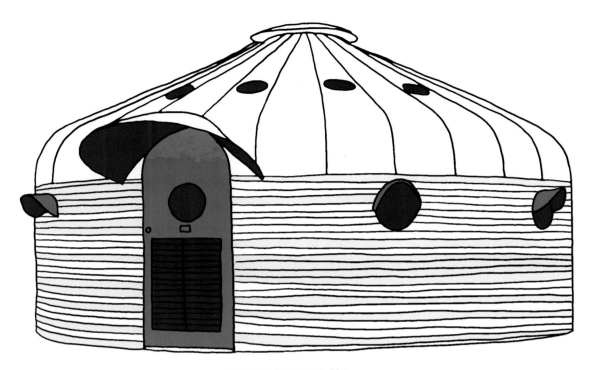

• A Dymaxion Dwelling Unit (DDU).

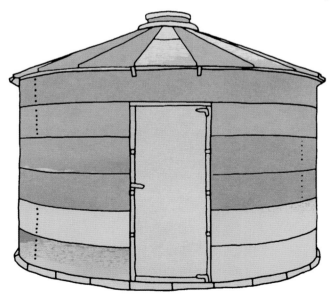

• A typical 1940's grain bin that Bucky thought could be modified into housing.

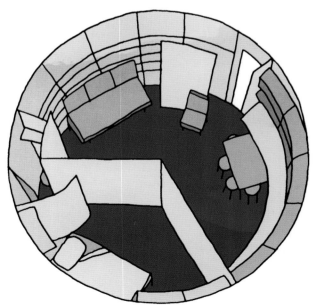

• A floor plan view to show how the DDU could be divided into rooms.

WICHITA HOUSE

In the mid 1940's, the end of World War II brought another opportunity for Bucky to propose a new version of his Dymaxion House. Many men would be returning to the U.S. from war and not only be in need of jobs, but they would also all be eager for housing as well. Bucky proposed to the military a new type of barracks, housing and a hospital. His design had evolved since the 4D house and Dymaxion Houses from previous years. The new design was circular, but still hung from a mast like his previous dwellings. It took advantage of new types of metal that had been developed. The military ordered two prototypes to be built, but not for military purposes. They saw the need for other housing possibilities that were pressing.

At the time a lot of airplane manufacturers were located in Witchita, Kansas. Kansas was a much harder target for enemies to locate and bomb being so far from the coast. Many skilled workers remained here building airplanes. The conditions were hard, and there was a housing shortage. The military saw a reason to start building Bucky's housing as a way to keep people in jobs and to provide housing for them at the same time. It was here that Bucky set up shop and started building a prototype of his new "Dymaxion Dwelling Machine" with Beech Aircraft facilities and workers.

The whole house was designed to fit into a tube that could be delivered to the site on a truck and assembled in a few days. Made mostly of stamped metal, it would weigh about as much as a car and cost about the same as well. The shell was held together with tension and compression. The weight was all transferred to the mast, which was set into the ground in a very relatively small foundation. There was a large vent on top that would move with the wind and provide ventilation. Bucky applied the knowledge of ship-building and design, combined with the materials and technology of airplane manufacturing, using them all to his advantage.

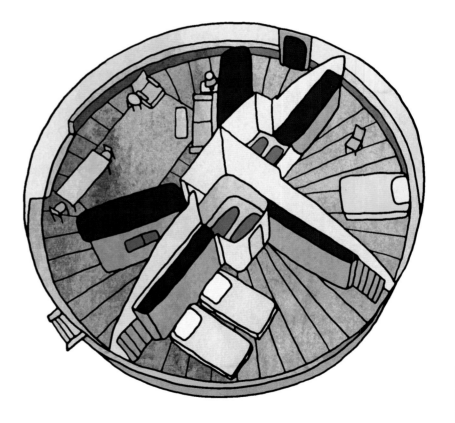

• Floor plan of the Dymaxion Dwelling Machine.

In 1945 a prototype of the Dymaxion Dwelling Machine was shown to the public. It was applauded and *Fortune Magazine* did an article on it. Thousands of people ordered a house. It seemed Bucky was well on his way to providing mass produced housing to the masses and, at the same time, making a small fortune. That couldn't be further from the truth. Bucky didn't think that the infrastructure to set up distribution and assembly would be ready, nor did the electricians and plumbers help at all. Workers refused to hook up any of the houses to the public services at the time. They were the only ones licensed to do so, and these new houses threatened their livelihood. All of this and Bucky was still not really even satisfied with the design as he continued working on it. Bucky predicted the design would take another seven years to complete.

Pressure built as investors hounded Bucky to speed up and get production going so everyone could make their fortune. Everyone was in it for the money, and Bucky didn't want to be a part of that. A frustrated Bucky walked away from the project and it went out of business. The prototype "Wichita House" was bought by a local man who put it on his land and lived in it with his family. It now belongs to the Henry Ford Museum, in Dearborn, Michigan, and is available for viewing to the general public.

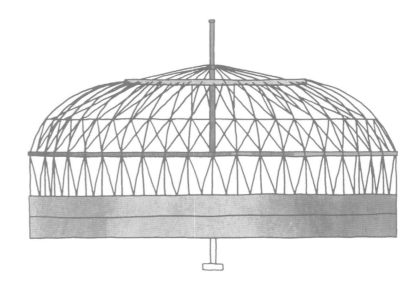

• The skelton was comprised of a series of tension rods not unlike spokes on a bicycle wheel. It would be sheathed in sheet metal and a large vent on top would rotate with the wind and circulate fresh air.

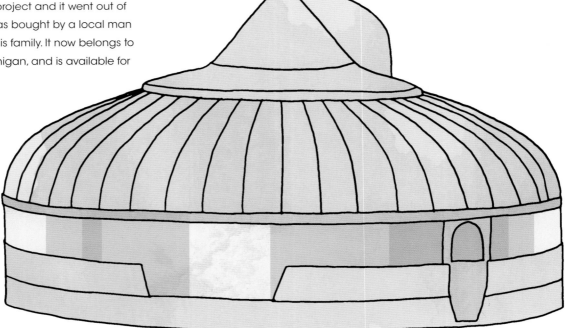

• The 36 foot diameter Wichita House was shipped in a tube and put together in a few days.

PROJECTING THE WORLD

Bucky had always been interested in the relationship of man on a global scale. He had done studies on the world's resources and how they are distributed. As air travel gained popularity, the world now seemed smaller. The traditional flat maps of the time were based on the Mercator Projection which distorted parts of the globe tremendously. Bucky designed a method, called the Dymaxion Projection method, in which a new map was devised.

An image of the earth is projected onto an icosahedron (a regular polyhedron with 20 identical equilateral triangular faces) using great circle grids. This shape can be unfolded into triangles that are flattened to reveal the whole planet with visually no distortion. It shows the interconnectedness of all of the continents and displays them as what Bucky called "one island earth". The triangles could be arranged in different ways to center on a certain part of the globe for strategic purposes.

In 1943, *Life Magazine* presented Bucky's map to the world in printed form. They printed it as a center spread with instructions on how to cut out and fold the map into a globe. He went on to patent the map in 1946 under the name The Dymaxion Airocean World Map. Bucky would continue studying and researching ways to divide a sphere and use that knowledge to develop practical applications and inventions. This research would help lead to the creation of the Geodesic Dome, Bucky's most well known invention.

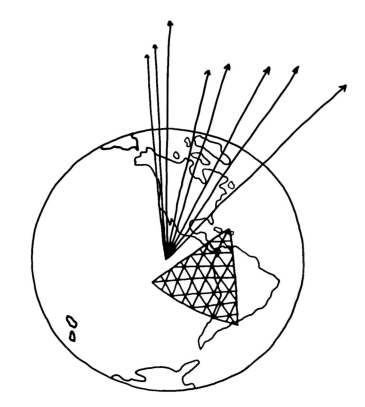

• Bucky's projection method translated the earth's surface to equilateral triangles that could then be flattened out to view the whole planet at once with little distortion.

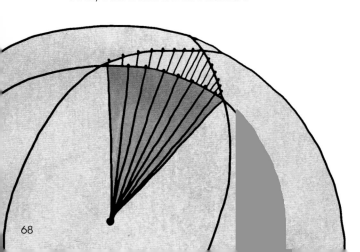

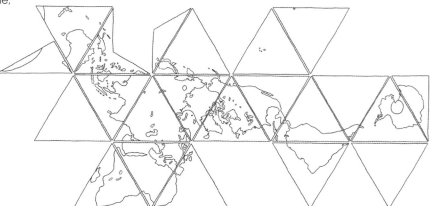

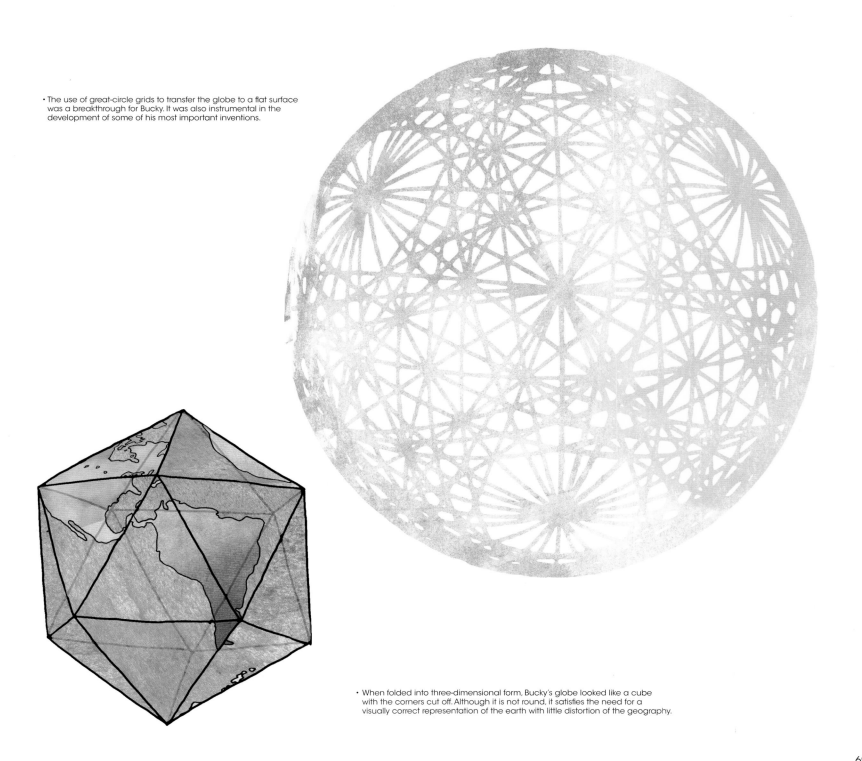

• The use of great-circle grids to transfer the globe to a flat surface was a breakthrough for Bucky. It was also instrumental in the development of some of his most important inventions.

• When folded into three-dimensional form, Bucky's globe looked like a cube with the corners cut off. Although it is not round, it satisfies the need for a visually correct representation of the earth with little distortion of the geography.

• The Dymaxion World Map's panels could be arranged
on a particular part of the world. Strategic air power wa
in the 1950's, particularly between the U.S. and the Sovi
one of the first people to realize the importance of the A
the fact that by air, the quickest way from the U.S.S.R. to
over this area.

time Bucky was developing the Dymaxion map, he
on what he called "Synergetics". After reading studies
n, Bucky explained that the energy of the universe is
is always conserved. In his two volume book, Syner-
s in the Geometry of Thinking, Bucky created a new
erse on his own terms. It combines science, poetry
to one work. Bucky defined synergy as "the behavior
not implicit in any of the behavioral characteristics of
f the system when those parts are considered only
ynergetics as the "exploratory strategy of starting with
opted "synergetics" as the name for the experimen-
he developed and demonstrated using numerous

on is that the entire universe is based on a tetra-
a pyramid with a triangular base. In this magnum
ibes why he thinks all physical and metaphysical
based on this form. Traditional geometric studies and
ased on the cube. Bucky argued that this was wrong,
on was the basic building block of nature. Traditional
ake the fourth dimension, time, into account, therefore
ational numbers, such as pi. With Synergetics, physi-
or more dimensional relationships could be visibly
or the first time.

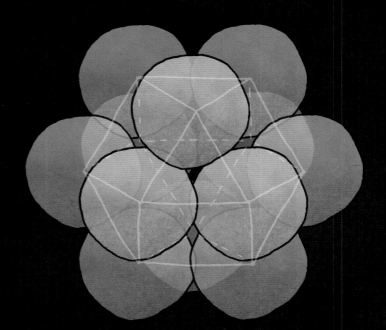

• From his studies with the "closest packing of spheres" around a
nucleus, Bucky developed a geometric model of a cuboctahedron.
In geometry, a cuboctahedron is a polyhedron with eight triangular
faces and six square faces.

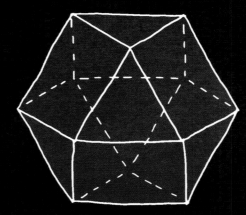

• Finding that this cuboctahedron is in perfect balance, he named it the
Vector Equilibirum or the "Dymaxion".

...y had essentially thrown out traditional teachings and found his ...geometry of nature. He began by studying the packing of spheres. ...h spheres are packed closely together in concentric layers around ...ntral sphere, certain regularities become apparent. It takes 12 ...res to make up the first layer, 42 for the next, and 92 for the third. ...y saw that this was the total number of naturally occurring chemi- ...elements, the 92nd in the periodic table of elements being ura- ...a. If you drew a line from the center of each of the 12 spheres, you ...what Fuller called the "Vector Equilibrium" or "Dymaxion".

Each line is the same length and everything is in perfect balance. Once you take out the middle sphere, you are left with a cubocta- hedron which has eight triangular faces and six square faces. Buc found an interesting lesson now because the pattern is no longer i state of equilibrium and the energy can reorganize into new patter Bucky called this transformation the "Jitterbug Transformation" after popular dance of the time.

• If a Vector Equilibrium starts rotating in on itself, it starts to make new patterns. Bucky named this the "Jitterbug Transformation".

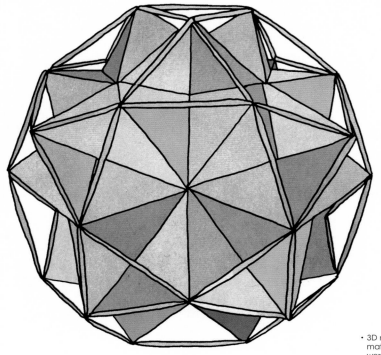

• 3D models of some of Bucky's geometric studies. Various materials such as paper, toothpicks, or even Venetian blinds were used to build his models.

Another structural principle Bucky studied was "Tensegrity", a term Bucky coined for tensional integrity. Tensegrity structures rely more on tension for its structural support than it does compression. You can think of it as components of rigid rods that do not touch each other, but are held together by cables. Bucky had already seen the potential in this years before while studying the wire wheel. This sort of structure would help spread the load bearing duties equally throughout the structure. These complete structures become rigid, but lightweight, for their overall mass. While teaching at Black Mountain College, Bucky's student Kenneth Snelson created the first model that demonstrated tensegrity.

• A tensegrity structure made of string and wooden dowels. While being structurally rigid, none of the dowels actually ever touch each other.

GEOSCOPE

Working with students at Cornell University in 1952, Bucky created what he called a Geoscope. This sphere was 21 feet in diameter and had a platform in the middle. The outside of the sphere was covered with a wire mesh that represented the earth's continents. The continents were placed so that if you stood inside the Geosphere, you could observe the universe and stars from different perspectives. This study was not, however, just to view the stars, but it was intended to view humans place in the universe at the same time.

• "Minni Earth" over the East River in New York.

MINNI EARTH

Bucky presented another Geoscope as an idea for helping students not lose focus when teaching about world planning. The purpose was to show statistical projections using millions of colored lights which would be hooked up to a computer. The lights could represent data collected on the computer and be rearranged for studying different areas of concern such as world resources and human migration.

In 1956 he proposed a 200 foot metal "Minni-Earth" version of this Geoscope to be placed outside of the United Nations building in New York City. This particular Geoscope could show data to the public and would be constantly updated. Bucky suggested using the Communist color red to illustrate population growth. As the public watches the wide expansion of the red lights, it now becomes a political tool at the same time as showing an ecological threat.

DOME IS HOME

During 1948 and 1949, Bucky spent time with students while teaching at both Black Mountain College in North Carolina and at the Institute of Design in Chicago. At Black Mountain he met and worked with Josef Albers and taught alongside William de Kooning. With the help of his trusted students, Bucky kept building models based on all of his studies.

From building small models based off his great circles out of venetian blinds, he continued to build larger dome structures. Bucky called these necklace domes because the structure was made of tubes with cables running through them. One of the necklace domes he designed with students as a project was called Skybreak. This particular dome could be stored in a container and then unfolded into a dome structure. It was then covered by a transparent skin, enabling the people inside to have a full view of the world, but still being out of the elements.

To coincide with the Skybreak dome, another assignment Bucky presented was the Standard of Living Package, The project was to design complete house furnishings for six people and to fit them into a standard container that could be pulled by a trailer. When the container was opened, the walls of the container folded down to become the floor of the living area.

• An early dome called "Skybreak" covering the "Standard of Living Package".

MONTREAL ARCTIC DOME

In February of 1949, the government invited Bucky to set up a necklace dome in the garden of the Pentagon. It could be erected quickly and folded down into a small package. People were starting to realize the benefits of these structures, and it gathered a lot of attention.

The next year, due to aluminum being rationed in the United States, Bucky and some of his former students went to Montreal, Canada, to build another prototype dome. This was the biggest one yet with a 50-foot diameter. This dome was to be used as an Arctic installation and utilized aluminum tubes with a weatherproof skin applied to the inside of the structure. The magazine *Architectural Forum* did an article on this dome, and its potential uses. The article included a profile of Bucky and some of his earlier work. Patents were soon filed and granted. Bucky was about to get a big break.

• The dome was light but extremely rigid so men could climb on it to fasten the interior cover to the aluminum tubing.

WOODS HOLE

In 1953, Bucky got his first freestanding geodesic commercial project. It was for a restaurant in Woods Hole, Massachusetts. An architect and restaurateur, Gunnar Peterson, hired him to design the structure, which would house a transparent dining room overlooking the ocean. It was a wood structure made of interlocking triangles covered with glass. At night you could view the stars from your table, but still be out of the elements.

Being such an unusual structure for the time and the location, it did not win all of the locals over. Also, being the first real dome for commercial use, it had its problems. The glass worked like an amplifier for all the music and conversation inside the structure.

With all of its glass windows, it also created a greenhouse effect, so it got really hot inside. It was eventually covered with white fiberglass to protect it from the sun, but then the diners couldn't see the stars anymore. It also was notorious for leaking. The restaurant was eventually closed, and the dome sat vacant for years. It was in disrepair for over 50 years. Developers bought the property, and the dome was to be torn down. The neighborhood rallied to save the dome, and for now, it has been marked for restoration, but still awaits its fate.

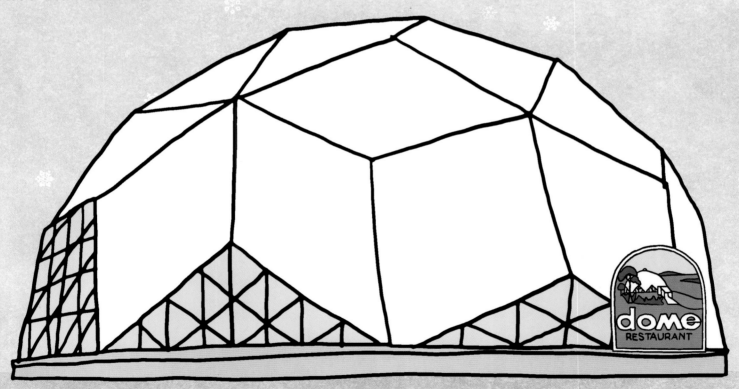

• Bucky's first commercial dome was for a small restaurant.

• A close up of the dome structure, covered in a plastic skin.

FORD ROTUNDA

Also In 1953, at the age of 58, Bucky got his first large-scale commercial request for one of his domes from none other than the Ford Motor Company. Bucky had been an earlier admirer of Henry Ford's industrial thinking and it seemed apropos. Although Henry Ford had recently passed away, he had requested a dome be built over their building called the Ford Rotunda. The building had originally been built for the Chicago World's Fair in 1933, but was moved and reassembled in Dearborn, Michigan. For the Ford Company's 50th anniversary, they wanted to enclose the open interior court.

Henry Ford II hired Bucky, and the dome was designed, built and installed just four months later. Its important to note that Bucky wasn't an architect. He did, however, have architects working with him. He didn't even have an office at the time. George Nelson, iconic modern furniture designer, lent him some space at his office. The dome was created using the octet truss (the structure he built out of peas and toothpicks in kindergarten) instead of the great circle method, because he wanted the dome to be much stronger. It was covered in a plastic skin that let light in, but kept out rain. Time magazine ran a four-page spread about the dome, and overnight everyone now wanted a dome.

• Once open to the elements, the dome fully enclosed the building, but still let light in.

• An octet truss structure.

• A typical radar station setup around the Arctic Circle.

RADOME

Having successfully gotten the attention of the military throughout the years, Bucky soon got a series of commissions from them under direction of the Defense Department of the United States government. They were setting up a radar early warning system they called the DEW (Distant Early Warning) line in the Arctic.

The DEW line was to protect the United States and Canada by warning of a possible attack by the Soviet Union during the Cold War. The system of radars stretched from Alaska to Greenland. Being in the Arctic, the radars needed to be protected from heavy wind, ice, and snow, but still allow their radar system to work without interference. They also needed to be lightweight for delivery and to be able to be set up quickly.

Bucky designed a fiberglass ¾ sphere dome called a Radome (combining "radar" and "dome") that fit all of these needs. The Radome was tested and approved. Geodesics, Inc. was formed after getting the contract from the military. From 1954 through 1956, Bucky would make the most money he ever made in his life off of this contract. His royalties were estimated to be in the millions of dollars.

In 1959, The Museum of Modern Art in New York City hosted a show in the sculpture garden called *Three Structures by Buckminster Fuller*. The exhibit included one of Bucky's Radomes, along with a Tensegrity Mast and an Octet Truss.

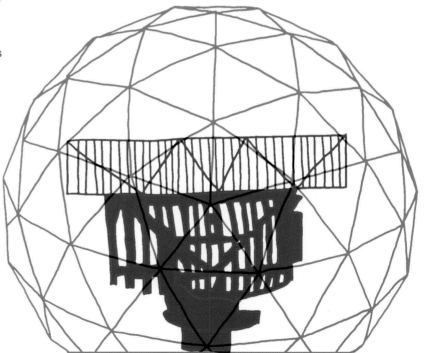

• An overlay of a Radom covering the protected radar.

U.S. PAVILLION

Bucky's domes had many advantages over traditional methods of building. They were ideal for quick or temporary structures. He would get many last minute requests to design and build domes with just a few months or even days to get them done. One of these instances happened when the U.S. State Department contacted Bucky's offices and wanted a dome for the 1956 Jeshyn International Fair in Kabul, Afghanistan. The U.S. Pavilion structure had to be designed, built, and installed in three months. It was to be air lifted and put together by local Afghan workers.

The dome was 100 feet in diameter and had an aluminum structure with a nylon tent that was attached to the frame from the inside. The components were easy to put together and to take down for compact shipping. After the fair in Afghanistan, this particular dome ended up being used all over the world to represent the United States at various gatherings. It went to places such as Tokyo, Burma, Bangkok, Peru, the Philippines, and Africa. This of course, gave Bucky even more recognition for his work.

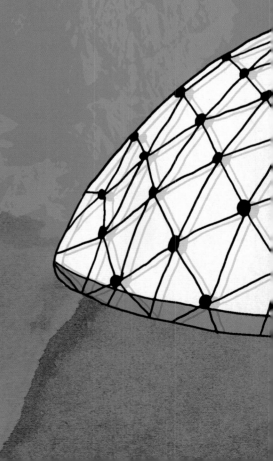

· With it's quick setup, Bucky's domes were
popular for use in remote locations where
shelter was needed in a hurry.

CARDBOARD DOME

Always experimenting with new materials and more efficient and cost effective ways of building structures, Bucky developed a design for a cardboard dome. Only two were made at the time, one for the military and one for an exhibition at the Triennale of Milan. The components could be shipped flat and erected into a three-dimensional form by folding the precut pieces of cardboard into triangles.

Lines printed on the cardboard pieces indicated where windows could be cut and folded into place. The domes were coated with a resin that hardened to help keep its structure and make it stronger. The resin would also make it waterproof. When the 42-foot dome was installed, it won the prestigious Gran Premio prize, gaining Bucky more press and interest in his domes.

• Holes could be cut anywhere for instant windows.

As the experiments continued for the military, a lightweight dome was developed for the U.S. Marine Corps that could be airlifted by a helicopter and dropped into place. The dome had a rigid outside structure, and a vinyl tent was attached from the inside, similar to the dome erected in Afghanistan. It was originally conceived to cover the very helicopters that were transporting it. It is reminiscent of the concept Bucky had of the zeppelin delivering the 4D house back in the 20's, except now it was a helicopter delivering the shelter which is now a dome.

In 1956, at the National Air Show in Philadelphia, the Marines demonstrated the lightweight capabilities of the 42-foot diameter dome by having a helicopter pick it up off of an aircraft carrier and fly it by the crowd.

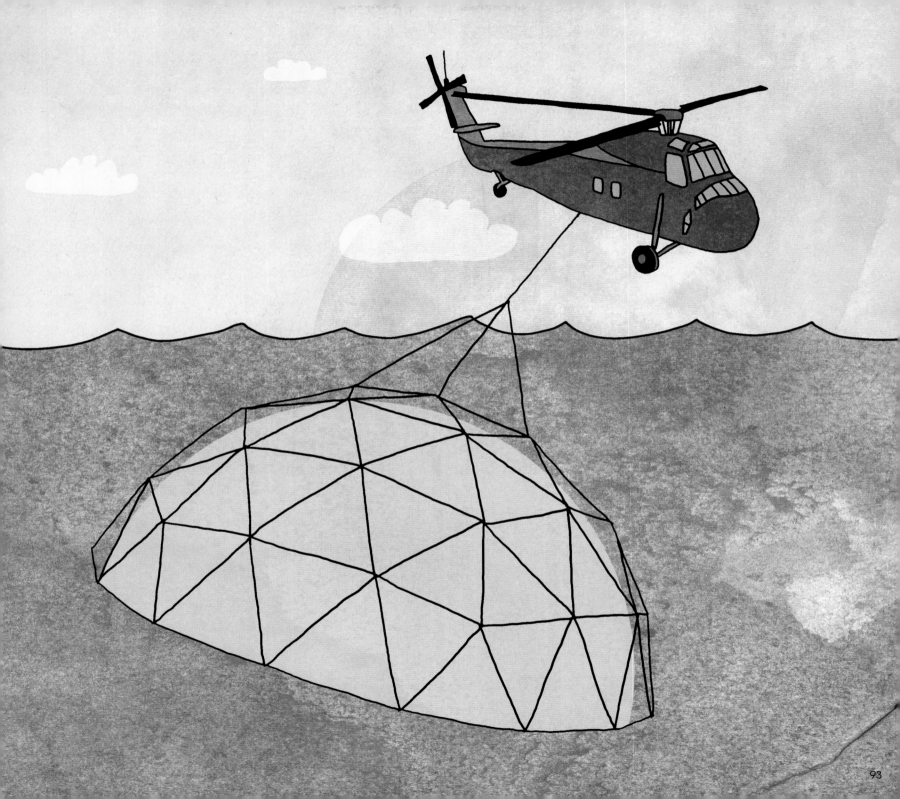

KAISER ALUMINUM DOME

Bucky and his team could not keep up with the demand for building domes if it had not been for the fact that he could license the designs to others and have them build it. That is why it was very important early on that the idea was patented. One of the first companies to license the geodesic dome was Kaiser Aluminum in 1956. Remember the guy Henry Kaiser that Bucky tried to design a car for back in the 40's, now he wanted to make domes out of aluminum, market, and sell them.

A former associate of Bucky's, Donald Richter, now worked for Kaiser Aluminum. Richter designed an aluminum dome to be built at Kaiser's Hawaiian Village property in Honolulu. All the parts were made in California and shipped to Hawaii. They built the dome from the "top down", meaning they built the top of the dome, moved it up a support structure, and then added the pieces below it until the whole dome was finished. This is the complete opposite of how most buildings are built, from the "ground up".

The speed at which this dome went up was mind-boggling. The day the dome started being built in Hawaii, Kaiser left San Francisco to watch its progress in person. By the time he got there, less than a day later, the dome was already built. There was a symphony concert in the dome that night with around 1800 guests in attendance.

the Kaiser Aluminum Dome

WOOD

MACHINE

BLACKSMITH

WELDING

WHEELS

RAMP

INCOMING →

INSPECTION

TRANSFER TABLE →

← OUTGOING

CONTROL TOWER

UNION TANK

• Interior plan of the rail car repair system.

UNION TANK CAR COMPANY

In 1958, a man named Dick Lehr was working on a new ingenious method of repairing oil tank rail cars for the Union Tank Car Company in Baton Rouge, Louisiana. He developed a circular carousel in which the cars could come in and be turned to each station, thereby increasing the rate at which these cars could be repaired. They needed a way to enclose this large space. Union Car Company soon became another licensee of Bucky's geodesic domes.

The span of this dome was extraordinary as the diameter would be over 400 feet, and the height would be over 130 feet. This would make it the largest span in the world. Bucky noted that the second largest cathedral in the world, the Cathedral of Seville, would easily fit inside the dome, but the dome would weigh less that 5% of what the cathedral weighed due to the type of construction and materials used. The Union Car Company would go on to build another dome in Illinois. These domes were widely publicized at the time, but unlike a lot of the domes built for the public or in densely populated areas, these domes were mainly out of site of the general public.

GLIMPSES OF THE UNITED STATES

The late 50's were the beginning of the Cold War between Russia and the United States. The United States had sought to promote its people and had been engaged in an exchange of policies with the U.S.S.R. They went into a cultural exchange agreement in which Russia would come to the United States and give us a little insight into their society and lifestyle. The U.S. would in turn do the same in Russia. In 1959, President Eisenhower and Russian leader Krushchev met and decide that a full presentation of life in each other's countries was in order.

The United States Information Agency hired George Nelson to design the exhibition that would be held in Moscow the next year. Nelson in turn went straight to Bucky for the design of the main encompassing structure. Being the late 50's, this was the pinnacle for mid-century modern design, and Nelson also hired none other than design icons Charles and Ray Eames to help him design the interior of the dome. Bucky would go on to design a golden anodized aluminum dome that would span about 200 feet.

The interior of the dome contained everything from space technology to kitchen appliances to a talking computer that would answer over 3000 questions. There was also a 360-degree Disney motion picture, which gave viewers 20-minute tours of cities in the United States. One of the most impactful presentations at the exhibition was a multiple screen installation by Charles and Ray Eames entitled "Glimpses of the United States" in which 7 giant 20 x 30 ft screens were suspended from the ceiling of Bucky's dome. On the screens, thousands of still images were projected and changed out quickly, creating an avalanche of information on the viewer. The images started out in outer space, slowly working down to aerial views of cities. It then panned all the way down to a typical United States citizen's daily habits, only actually showing people at the very end. Almost 3 million people would view the presentation. The Russians liked the dome so much that they ended up buying it.

• George Nelson hired Bucky to design the Moscow dome. The video installation was done by Charles & Ray Eames.

• Left: George Nelson
Middle: Buckminster Fuller
Right: Charles Eames

99

HOLLYWOOD HILLS DOME

After studying at the USC School of Architecture and hearing lectures by Bucky in Los Angeles, Bernard Judge decided he wanted to build a geodesic dome to live in. Not only would this be the first large scale dome for residential use, but he was to build it on a steep slope. Judge would use many donated materials and build it himself, alongside other architect students. He called his dome the "Triponet". This being 1958, dealing with the building department in Los Angeles and using new types of materials would be challenging.

The process of building it would take over four years. During this process, none other than *Life Magazine* stumbled upon Judge building his dome. Some of the photos taken were later published in an issue of the magazine about Los Angeles in 1960. After the article came out, Bucky himself even stopped by to see how the dome was coming along.

The dome was covered mostly in a transparent Mylar and on top had an additional silvery sunshade to help with the LA heat by reflecting the sun. The living structures inside were built totally independent of the dome itself. The rooms were essentially floating decks within the dome, a concept similar to Bucky's Skybreak and the autonomous living package.

After seeing the *Life Magazine* spread, the famous architectural photographer, Julius Shulman, went to see and photograph the house. These photos would be published numerous times. From these would come one of Shulman's most iconic shots. The house was featured on the cover of *The Los Angeles Times Home Magazine* in 1962. Due to a prior agreement, Judge had to give up the home to an investor after living in it for a year. Bernard Judge is now a celebrated and accomplished architect. The "Triponent" dome house has since been torn down.

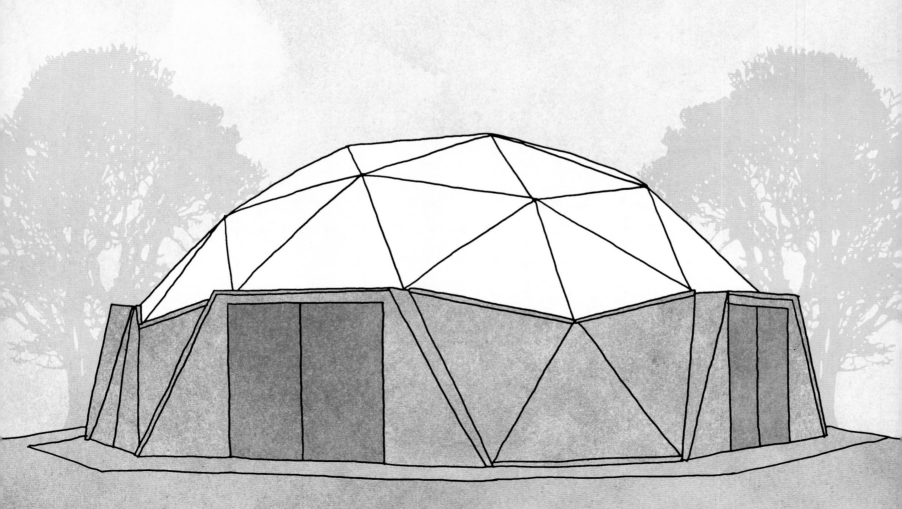

FULLER DOME HOME

In 1959, Bucky received and offer to be the Research Officer for Southern Illinois University in Carbondale, Illinois, which he accepted. He received a staff and a building to continue his research, but was still allowed to travel and give his lectures, which he did often. He and Anne moved to Carbondale and built their first house. 407 South Forest Avenue would become the home to the world's most important dome structure.

Bucky had licensed the geodesic dome patents to the Pease Wood-working Company in Hamilton, Ohio. Bucky and Anne's house would be the first residential application from this company. Built mostly out of wood and modest in size, at just over 1200 square feet, the house included a large living area where Bucky would invite students over and give lectures. It contained two bathrooms and one bedroom and a balcony library. The house took only seven hours to build and cost about eight thousand dollars. This house epitomized the "doing more with less" philosophy. It is considered the prototype for all other dome homes that followed.

Bucky and Anne lived here for 11 years, and although he traveled constantly, some of his most important work was produced while he lived in Carbondale. This was the only dome that Bucky ever lived in, and the only home he ever owned.

In 2006, the Fuller Dome Home, as many call it, was added to the National Registry of Historic Places. Over the years the house has been in disrepair. The house is currently undergoing renovation so that future generations can enjoy and find inspiration from it.

• Bucky and Anne's own house, a modest 1200 sq ft wood dome.

CLIMATRON®

In 1959, the Missouri Botanical Garden was celebrating its 100th year of existence. Many of the buildings in the garden were outdated and in disrepair. The Board of Trustees wanted something modern and dramatic to replace these buildings.

In considering all aspects of the reconstruction problem, Mr. Eugene Mackey of the Architect firm of Murphy & Mackey, which had been retained to do the planning of the new greenhouse, decided that a geodesic dome, as invented by Mr. Buckminster Fuller, would best fulfill all requirements. Open to the public in 1960, The Climatron® was the first geodesic dome to be used as a greenhouse. The greenhouse was named Climatron to emphasize the climate-controlled environment, as it is completely air-conditioned.

The structure is made up of aluminum rods and tubes and was covered with plexi-glass panels that later were replaced by actual glass. The dome houses more than 2,800 plants, and the tropical rain forest inside the dome spans over half an acre. The height of the dome is approximately 70 ft. Within the dome viewers can experience waterfalls, cliffs, aquariums, and a bridge to view the forest canopy.

In 1976, the Climatron dome was named one the 100 most significant architectural achievments in United States History.

• Interior of the Climatron.

• Exterior of the Climatron at night.

MONTREAL "EXPO 67"

In 1964 Bucky was asked by the United States Information Agency to submit a design for the U.S. Pavilion at Montreal's "Expo 67". The agency wanted something to make a statement on the scale of what the Eiffel Tower did at the Paris World's Fair in 1889. Bucky would be up against many other firms for the job. Bucky had been working with architect Shoji Sadao who was a mutual friend of Isamu Noguchi. They decided to form Fuller and Sadao Inc. to pitch their ideas. They got the job.

Their first proposal was for a space frame structure that housed a balcony where the viewer could look down on Bucky's Dymaxion Map the size of a football field. On the map would be lights representing energy, food resources and economic trends that were interactive based on a viewer's selection of criteria. This brought global political powers to an individual, and this person could now see how choices that are made effect someone on the other side of the planet. It would show how to use all the resources the earth has to offer in the most beneficial distribution to mankind. Bucky would call this experience the "World Game". To create this interactive map, a huge database would need to be made with inventory of all the world resources and statistics. A large mainframe computer would be needed to crunch all this data. It would also need to monitor the world's activities and be updated in real time. None of this technology existed at the time. This proposal was not well received.

• Opposite: The interior to the Montreal Expo dome.

It is important to note that while the "World Game" was not implemented by the agency, it was one of Bucky's most enduring philosophies. In 1972 he founded the World Game Institute. Bucky continued to teach and host seminars where simulations were created. Participants could help create solutions to the world's problems, such as overpopulation and the uneven distribution of resources.

Currently the Buckminster Fuller Institute (bfi. org) hosts the Buckmister Fuller Challenge, which is an annual international design challenge that awards money to support the development and implementation of a strategy that has significant potential to solve humanity's most pressing problems.

• World Game interactive map proposal.

After the World Game structure was rejected, a dome was proposed and agreed too. The dome designed was a ¾ sphere, 200 feet high and 250 ft wide. This would be the largest dome to date by volume. Making it a ¾ sphere would drastically increase the ceiling height, bringing a dramatic effect once inside the space. The dome was covered in a transparent acrylic skin so it was essentially open to the sky and the sun. To provide protection from the sun, a series of computer-controlled shades on certain individual acrylic panels of the sphere would open and close depending on the sun's position in the sky.

A group called the Cambridge Seven designed the interior. It consisted of multilevel platforms in which a viewer could experience the space from different perspectives. On display throughout were some of the United States Space program rockets and other space related structures. Art from some of the U.S.'s most well-known artists was on display. To add to the dramatic experience of the dome, there was an elevated, open-air monorail that went all around the expo. When it got to the dome it would literally go right through it. The feeling would be as if you are flying through the space.

Millions of people from around the world would visit the Expo and see the dome. Bucky received instant praise for the building. It was a huge success and was given an award from the American Institute of Architects in 1968. People would call it Bucky's "Taj Mahal", as it was his masterpiece. Bucky dedicated the building to his wife Anne. In 1969, continuing to ride his wave of recognition, Bucky was nominated for a Nobel Peace Prize. The same year the American Humanist Association named him humanist of the year.

In 1976 the dome was accidentally set ablaze while a welder was repairing some seams in the panels. The whole exterior was destroyed, but the frame remained intact. It still stands today as a museum called the Montreal Biosphere.

• The dome housing the United States Pavillion at Montreal's "Expo 67".

FLY'S EYE

Bucky continued to develop different dome systems. The "Fly's Eye" dome was a convergence of all his ideas as he still tried to find the perfect solution for an autonomous dwelling machine. This dome could be made out of lightweight material and be air delivered. It would weigh and cost about the same as an automobile. It's design made it easy to manufacture and build, thus reducing costs. Although light in weight, its structure could withstand heavy snow loads or even earthquakes and hurricanes.

It got its name from the circular windows that were created resembling the compound lenses of a fly. The circular openings could also be filled in with venting or solar panels.

Although the patents for this dome were filed in 1961, it was never fully realized, although a few prototypes were built out of fiberglass. A 50 foot Fly's Eye dome was built and exhibited in 1981, a full 20 years after its patent filing.

DOME LEGACY

Using principles of design employed by nature, Bucky was able to invent the geodesic dome. He saw that nature always worked in the most economical way, doing more with less. This came from years of studying geometry, building models, and continually trying to improve on past ideas. The big breakthrough came from studying the earth itself and looking at the shortest distance between two points on a globe. By disecting the globe with great circle grids, he came up with a building structure that mimicked nature's efficiency and had the greatest strength with least amount of materials. A structure that is stronger than its individual parts.

Bucky's goal was to make housing affordable and easy for everyone. The geodesic dome was a far more efficient way to enclose space than a traditional house. The sphere, in which it tries to approximate, has the minimum surface area to inclose the maximum volume. Geodesic domes are lightweight and easy to assemble. They can be manufactured at a factory and then shipped to the site to be put together. This made it one of the first "pre-fab" housing solutions.

Due to the domes shape, it also has the least heat gain or loss compared with traditional structures of the same floor area, therefore making it more energy efficient. The shape is also more aerodynamic, as wind goes over and around it, rather than hitting directly against the walls of a vertical structure. This makes it less vunerable to violent storms.

Domes do not come without problems. One of which is adding on space. You cannot add a second story. Also on small lots, it would be hard to put a dome when it makes more sense to build up to hold the maximum amount of people for that space. Cabinets cannot be directly hung from the walls like in traditional houses, and windows can be tricky to deal with. Acoustics in a dome are much different, and they can cause annoying echoes.

Although there are domes in almost every country in the world, they did not entirely catch on with the general public. They have been popular for larger, more commerical structures, such as stadiums or expos. Most people are probably aware of geodesic structures such as Epcot Center or the Tacoma Dome in Washington State. While plenty of people built and still live in geodesic domes, they were mostly seen as more of a novelty. They are definitely not suited for everyone's taste. Many people view them as futuristic, or think they all look alike. Also, having been popular with the counterculture of the 60's and 70's, some people think only "hippies" would live in them. Many of the domes built by people with "do-it-yourself" instructions were built poorly, leaked and gave domes a bad name.

While Bucky is synonomous with geodesic domes, he wasn't actually the first person to build one. The first documented use of a geodesic dome was in 1922 by Walter Bauersfeld. Bauersfeld's dome was used for a planetarium in Germany. Bauersfeld never applied for a patent, nor ever used a geodesic structure again. He apparently didn't even know what he had! It is also important to note that Bucky was not an architect and had many people help develop and create his ideas. Bucky held the patent on the dome, so when others designed a dome, he would get a license fee, but sometimes his name was never mentioned. The Epcot Center's geodesic sphere, built in 1982, is called "Spaceship Earth", but when it was revealed to the public, there was no mention of Bucky at all.

The geodesic dome was the only new type of structure to be discovered and developed in the previous 2000 years. Although there are hundreds of thousands of Bucky's domes throughout the world, the true measure of success does not lie in the numbers. The geodesic dome is a testament to Bucky's belief that innovative thinking, when applied to a good purpose, can make the world work better for all of us.

· Floating Tetrahedral City: Design by Bucky and Shoji Sadao.
Proposed for the San Francisco Bay in 1965.

THINKING BIG

Always thinking of how to manage housing for a large number of people, at a cost effective manner, Bucky proposed a series of large scale cities. These cities were self contained and had minimal impact on the environment because they were either located on the water, in the air, or even submerged in the ocean.

After his invention of the geodesic dome, he even proposed that existing cities could be enclosed by giant domes protecting the city from the elements. One of his proposals was the Dome over Manhattan, which would protect people from the hot Summer months and save an enormous amount of engery in the Winter by using solar heat to heat the city instead of trying to heat individual buildings.

Bucky knew his ideas may not be possible, at least in his lifetime, but he did start a dialogue about population expansion and the environmental impact of building housing for all those people.

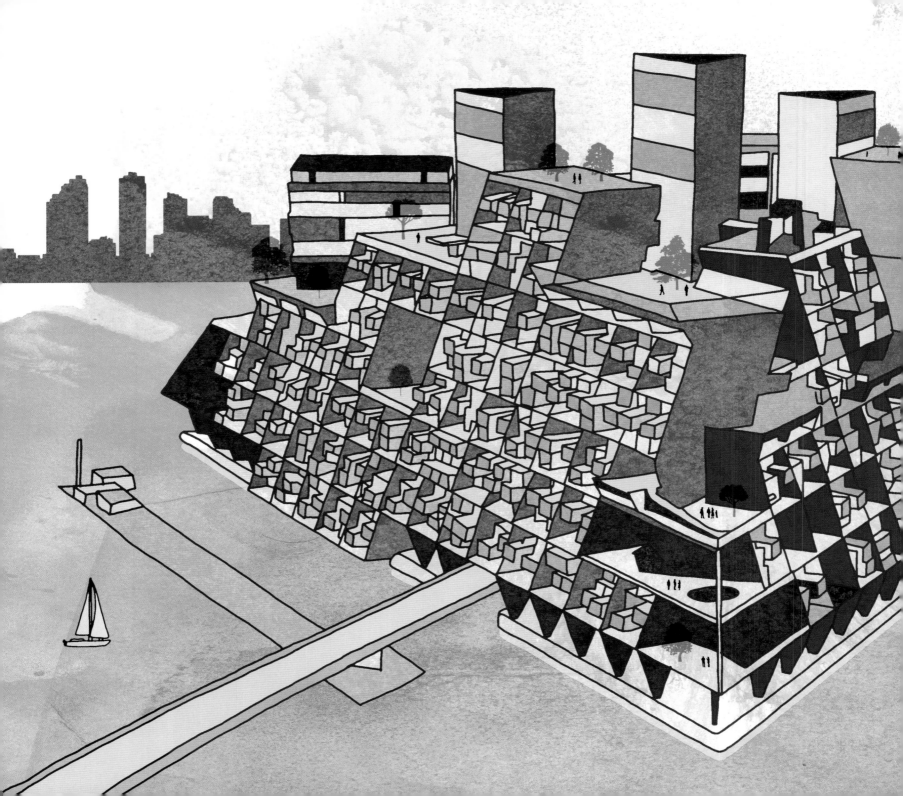

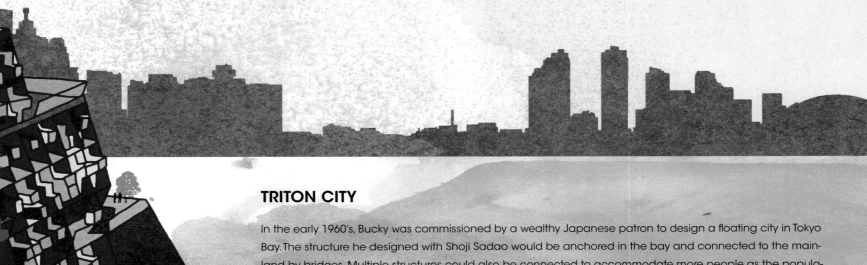

TRITON CITY

In the early 1960's, Bucky was commissioned by a wealthy Japanese patron to design a floating city in Tokyo Bay. The structure he designed with Shoji Sadao would be anchored in the bay and connected to the mainland by bridges. Multiple structures could also be connected to accommodate more people as the population grew. Bucky thought by not having to buy land on which to build a structure, that someone at the poverty level could afford this housing.

Bucky decided upon a design based on the tetrahedron, which has the most surface space and the least volume. He did this in order to maximize the shape for inhabitants to all have outdoor balconies with views. The design would also prevent fatalities if anyone happened to fall off a balcony, as the next level extends out from below. The building was resistant to tsunamis and desalinated the water for consumption. The buildings would not only provide housing, but also provide outdoor space, shopping, entertainment, and schools. The population of these floating cities could be upwards of 5000 people.

In 1966, the patron of the project died and the project halted. The United States Department of Housing (HUD) asked Bucky to continue developing his study for possible use in the U.S.A. After building models and doing a cost analysis for building the city, it was sent to the Secretary of the Navy where it was analyzed for its floating capabilities. After further positive studies, the city of Baltimore became interested in building this city in Chesapeake Bay, but due to several factors, including the shifting politics of the day, the project never came to fruition.

• "Triton City", a city on water able to increase in scale by connecting floating structures.

CLOUD NINE

Another proposal of high-density housing not using land was a floating geodesic sphere Bucky called "Cloud Nine". These structures would be up to a mile wide and could be tethered in place or have the ability to migrate in order to avoid weather, or just to see the world. These structures could be deployed to provide emergency housing after a natural disaster.

As these spheres grow in size, they also become stronger by means of distributing the stress over the surface. Bucky suggested that if the air inside one of these structures were heated by a mere 1-degree higher than the outside temperature they would simply float like a balloon. The air could be heated by either solar or the human energy inside the sphere or a combination of both.

OLD MAN RIVER'S CITY

One of Bucky's most well known large-scale structure designs is the Old Man River's City project in East St. Louis, Missouri, in the early 1970's. The name was taken from an old song about the people who lived on the banks of the Mississippi River in that area at the time of cotton trading. Bucky was asked to help in designing a solution to help the poverty stricken neighborhood with better and more affordable housing. Bucky, along with a team of architects including James Fitzgibbon, who was Washington University's Professor of Architecture, created a beautiful all encompassing design. The city consisted of what they called a "moon crater cone" with a geodesic dome cover.

The cone was terraced on both the outer portion and the inner portion. The shape somewhat resembles a low rising cinder cone volcano. The outer portion would be living quarters separated by trees and greenery for privacy. All units had views because of the way the cone was terraced up the side. None of the units need roofs because of the giant dome covering the whole structure! It would be like living in an outdoor garden. The interior of the cone would be common space for parks, tennis courts, supermarkets, and outdoor theatres, etc. On the interior ground level, the amount of space would be about the size of four football fields combined.

The dome itself would be a quarter-sphere translucent dome that would protect the city from the weather and help regulate comfortable conditions underneath. It was like a giant umbrella 1000 feet in the sky. The dome would collect water runoff to store for fire sprinklers and to be sent through a system for purification. The width of the city (and dome) was about a mile in diameter. The city could accommodate up to 125,000 residents.

The Old Man River's City project received a lot of attention and government money was allocated to help build it. Bucky warned the community of restrictions that would be imposed if they accepted the money. The government allocation was refused, and the community of East St. Louis, Missouri started the financing on their own. To this date no ground has been broken but as late as 1995 there was still an organized effort to realize the project.

· The width of this proposed dome was almost a mile wide.

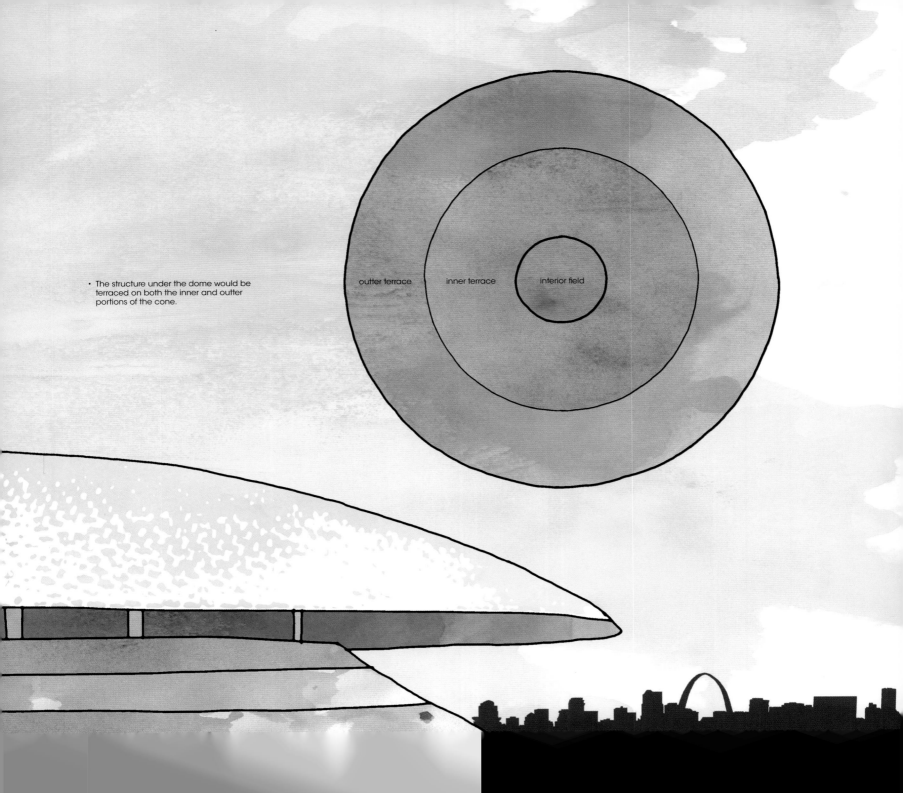

- The structure under the dome would be terraced on both the inner and outter portions of the cone.

outter terrace   inner terrace   interior field

BACK TO THE SEA

As we know, Bucky was a lover of the sea and of boats and being on the water. In 1970, he was granted a patent on a type of boat that he had been developing for many years. Rowboats are long and skinny and can tip over easily. Bucky designed a rowboat for a single person that was more stable than the types of rowboats of the time. He called it the Rowing Needles, also known as a Watercraft.

The Rowing Needles used a catamaran like design to add stability. The rower would sit on a platform that spanned in between the "needles". The patent would show how multiple Rowing Needles could be combined to make the craft a multi-oared boat. Only four prototypes were built. The first were built of aluminum, and subsequent prototypes experimented with other materials. The craft was never put into production.

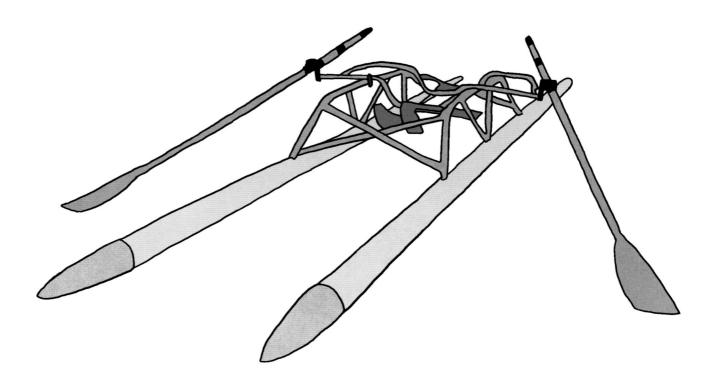

- Variations of the Rowing Needles were designed
  including a sailboat version.

CALL ME TRIMTAB

A trim tab is a small surface on a larger control surface, such as a rudder on a ship. The trim tab basically controls the rudder, which in turn controls the movement of the entire ship. Bucky used this as a metaphor for himself, saying that an individual can be a trim tab, one little man changing the course of the world.

In 1972, Bucky and his wife Anne moved to Philadelphia where he had accepted a professorship. He continued lecturing and had many exhibits of his work. In 1975 he recorded a series of lectures on his life's work. He called this series "Everything I Know", and it is approximately 42 hours in length. Also, in the same year, his book, Synergetics, was published. Synergetics is the culmination of decades of exploration into what he called "synergetic geometry". He would go on to publish Synergetics 2 in 1979.

Anne and Bucky moved to Los Angeles in 1980 to be closer to their daughter Allegra. Bucky still maintained his Philadelphia office and continued his busy lecturing schedule, which he maintained for years.

Only months after receiving the Medal of Freedom from President Ronald Reagan, Buckminster Fuller passed away on July 1, 1983, in Los Angeles, CA. His wife Anne Fuller died just 36 hours later. "CALL ME TRIMTAB" is etched on his gravestone.

By the time he died, Bucky had registered 28 patents, written 28 books, traveled around the globe 57 times and received 47 honorary doctorates. He won numerous awards including a Gold Medal of the American Institute of Architects, a Gold Medal of the Royal Institute of British Architects, and his 1969 nomination for the Nobel Peace Prize. His most well known design, the geodesic dome, is estimated to have been produced over 300,000 times around the world.

Bucky did not limit himself to one field but, worked as a "comprehensive anticipatory design scientist" to solve global problems surrounding housing, shelter, transportation, education, energy, ecological destruction, and poverty. He blurred the lines between art and science. He was living proof that fostering creativity can lead to scientific breakthroughs.

During his life Fuller worked with and influenced a diverse group of people from poets and artists to architects and scientists. Fuller ignited inspiration everywhere he went, delivering more than 2,000 lectures at 500 universities and colleges. He would present an average of 150 "thinking out loud" sessions every year.

Since his death the relevance of his radical inventions and proposals are still being realized and, as Bucky preached, it is now becoming clear that the world's resources are not infinite. In 1996, a Nobel Prize in Chemistry was given to a group of scientists who discovered a molecule that resembled Bucky's geodesic spheres. They named it the Buckminsterfullerene, or "buckyball" in homage to Bucky.

Bucky's true impact on the world can be found in his continued influence upon generations of designers, architects, scientists, and artists working to create a more sustainable planet.

THE DYMAXION CHRONOFILE

Way back in 1917, Buckminster Fuller began to catalog his life. He had already compiled scrapbooks of items that interested him starting back in 1907. All of his personal documents including sketches, newspaper clippings, itineraries, and other important papers were put into chronological order in 15-minute increments. This collection is called the "Dymaxion Chronofile".

Bucky intended for this collection to be a sort of case study of his life in relationship with technological advancement of society. He referred to himself as "Guinea Pig B", a human experiement. This experiment needed to be documented for others to one day study. He himself would often look back into his records, noticing trends of innovation and social change.

This gigantic scrapbook was bound into leather volumes, and when he died in 1983, they took up about 270 linear feet.

Given the enormity of this collection, he is probably the most documented person of the 20th century. The Chronofile, along with other photographs, physical models, artifacts, and video tapes belong to the Buckminster Fuller Archive at Stanford University.

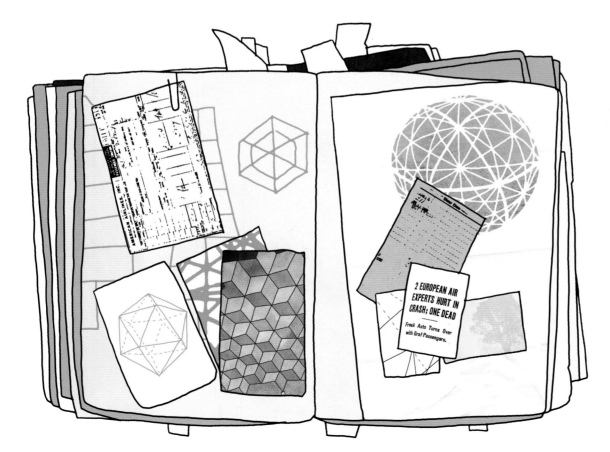

2 EUROPEAN AIR
EXPERTS HURT IN
CRASH; ONE DEAD

*Freak Auto Turns Over
with Graf Passengers.*

• Bucky cataloged his life including postcards,
  photographs, sketches, and even receipts.

"Whether it is to be Utopia or Oblivion will be a touch-and-go relay race right up to the final moment... Humanity is in 'final exam' as to whether or not it qualifies for continuance in Universe..."

R. Buckminster Fuller

R. Buckminster Fuller U.S.A Patents

Date of Granting	Invention	USA Patent No.
June 28, 1927	Stockade: Building Structure	1,633,702
July 5, 1927	Stockade: Pneumatic Forming Process	1,634,900
(April 1, 1928)	4D House*	(1,793)
December 7. 1937	Dymaxion Car	2,101,057
November 5, 1940	Dymaxion Bathroom	2,220,482
March 7, 1944	Dymaxion Deployment Unit (sheet)	2,343,764
June 13, 1944	Dymaxion Deployment Unit (frame)	2,351,419
January 29, 1946	Dymaxion Map	2,393,676
June 29, 1954	Geodesic Dome	2,682,235
April 14, 1959	Paperboard Dome	2,881,717
September 22, 1959	Plydome	2,905,113
November 24, 1959	Catenary (Geodesic Tent)	2,914,074
May 30, 1961	Octet Truss	2,986,241
November 13, 1962	Tensegrity	3,063,521
March 12, 1963	Submarisle (Undersea Island)	3,080,583
July 7, 1964	Aspension (Suspension Building)	3,139,957
August 3, 1965	Monohex (Geodesic Structures)	3,197,927
August 31, 1965	Laminar Dome	3,203,144
(March 17, 1965)	Octa Spinner*	(349,021)
November 28, 1967	Star Tensegrity (Octahedral Truss)	3,354,591
August 17, 1970	Rowing Needles	3,524,422
May 14, 1974	Geodesic Hexa-Pent	3,810,336
February 4, 1975	Floatable Breakwater	3,863,455
February 18, 1975	Non-symmetrical Tensegrity	3,866,366
January 30, 1979	Floating Breakwater	4,136,994
June 17, 1980	Tensegrity Truss	4,207,715
March 22, 1983	Hanging Storage Shelf Unit	4,377,114

*Application abandoned by Buckminster Fuller leaving prior art evidence in patent office files.

Selected Bibliography
Books by R. Buckminster Fuller

4D TIMELOCK. (Privately published 1928, Chicago, Illinois; 200 copies) Biotechnic Press, Lama Foundation, Albuquerque, New Mexico (1,000 copy edition). c1929, 1970. c1972, paperback.

NINE CHAINS TO THE MOON. J. B. Lippincott Company, Philadelphia, New York, London, Toronto c 1938, hardback; republished Doubleday & Company, Inc., Garden City, New York. c1963, paperback.

THE DYMAXION WORLD OF BUCKMINSTER FULLER. With Robert W. Marks. Anchor Press, Doubleday & Company, Inc., Garden City, New York. c1960, paperback.

UNTITLED EPIC POEM ON THE HISTORY OF INDUSTRTIALIZATION. Simon & Schuster, New York. c1962, hardback, paperback.

EDUCATION AUTOMATION. Doubleday & Company, Inc., Garden City, New York. c1963, paperback.

IDEAS AND INTEGRITIES. Prentice Hall, Englewood Cliffs, New Jersey. c1963, hardback; Collier, Macmillan, Toronto, Canada. c1963, paperback.

NO MORE SECONDHAND GOD. Doubleday & Company, Inc., Garden City, New York. c1963, paperback.

OPERATING MANUAL FOR SPACESHIP EARTH. E.P. Dutton & Co., New York. c1963, 1971, paperback.

WHAT I HAVE LEARNED. RBF's chapter—"How Little I Know." Simon & Schuster, New York. c1968, hardback.

UTOPIA OR OBLIVION. Bantam Books, New York. c1969, paperback.

THE BUCKMINSTER FULLER READER. Edited by James Meller. Jonathan Cape, UK., London. c1970, hardback. Penguin Books, Ltd., Middlesex, England. c1970, paperback. (available only in England).

I SEEM TO BE A VERB. With Jerome Agel and Quentin Fiore. Bantam Books, New York. c1970, paperback.

INTUITION. Anchor Press, Doubleday & Company, Inc., Garden City, New York. c1970 paperback; Impact Publishers, San Luis Obispo, California, paperback.

BUCKMINSTER FULLER TO CHILDREN OF EARTH. Text by Fuller, compiled and photographed by Cam Smith. Doubleday & Company, Inc., Garden City, New York. c1972, paperback.

EARTH, INC. Anchor Press, Doubleday & Company, Inc., Garden City, New York. c1973, paperback.

SYNERGETICS: EXPLORATIONS IN THE GEOMETRY OF THINKING. In collaboration with E.J. Applewhite. Introduction and contribution by Arthur L. Loeb. Macmillan Publishing Company, Inc., New York. c1975, hardback; paperback.

TETRASCROLL. Limited Edition. Universal Limited Art Editions, West Islip, New York. c1976. TETRASCROLL, St. Martin's Press, New York. c1975 and 1982, hardback.

AND IT CAME TO PASS – NOT TO STAY. Macmillan Publishing Company, Inc., New York. c1976, hardback.

SYNERGETICS 2: FURTHER EXPLORATIONS IN THE GEOMETRY OF THINKING. In collaboration with E.J. Applewhite. Macmillan Publishing Company, Inc., New York City, New York, c1979, hardback; paperback.

R. BUCKMINSTER FULLER ON EDUCATION. Edited by Robert Kahn and Peter Wagschal, University of Massachusetts Press Amherst, MA. c1979, hardback; paperback.

SYNERGETICS FOLIO: A COLLECTION OF TEN POSTERS WITH INTRODUCTORY ESSAY BY BUCKMINSTER FULLER. Privately printed. c1979, soft cover.

BUCKMINSTER FULLER SKETCHBOOK. University City Science Center, Philadelphia, c1981, paperback.

CRITICAL PATH. With Kiyoshi Kuromiya, adjuvant. St. Martin's Press, New York City, New York. c1980, hardback; paperback.

GRUNCH OF GIANTS. St. Martin's Press, New York. c 1983, hardback.

HUMANS IN UNIVERSE. With Anwar Dil, Mouton, New York. c.1983, hardback.

COSMOGRAPHY, A POSTHUMOUS SCENARIO FOR THE FUTURE OF HUMANITY. With Kiyoshi Kuromiya, adjuvant. Macmillan Publishing Company, New York c.1992, hardback.

Books about R. Buckminster Fuller

R. BUCKMINSTER FULLER. by John McHale. George Brazillier, Inc., New York. c1962, hardback.

BUCKY—A GUIDED TOUR OF BUCKMINSTER FULLER. by Hugh Kenner. William Morrow & Company, Inc., New York. c1973, hardback; paperback.

BUCKMINSTER FULLER AT HOME IN THE UNIVERSE. by Alden Hatch. Crown Publishers, New York. c1974, hardback.

COSMIC FISHING: AN ACCOUNT OF WRITING SYNERGETICS WITH BUCKMINSTER FULLER. by E.J. Applewhite. Macmillan Publishing Company, Inc., New York, c1977 hardback.

PILOT FOR SPACESHIP EARTH. by Athena V. Lord. Macmillan Publishing Company, Inc., New York. c1978, hardback.

BUCKMINSTER FULLER: AN AUTOBIOGRAPHICAL MONOLOGUE/SCENARIO. by Robert Snyder. St. Martin's Press, New York. c1980, hardback.

INVENTIONS, THE PATENTED WORKS OF R. BUCKMINSTER FULLER. St. Martin's Press, New York; c.1983

BUCKMINSTER FULLER'S UNIVERSE. by Lloyd Steven Sieden. Perseus Publishing, Cambridge, Massachusetts, c.1989

BUCKY WORKS: BUCKMINSTER FULLER'S IDEAS FOR TODAY. by J. Baldwin. John Wiley & Sons, New York, c. 1996, Hardback; 1997, Paperback.

A FULLER EXPLANATION: The Synergetic Geometry of R. Buckminster Fuller. by Amy C. Edmondson. Van Nostrand Reinhold, New York, c.1987.

YOUR PRIVATE SKY: R. BUCKMINSTER FULLER: THE ART OF DESIGN SCIENCE. Edited by Joachim Krausse, Claude Lichtenstein (Editor), Zurich Museum of Design. Lars Muller Publishers, c.2001. Hardcover.

YOUR PRIVATE SKY: R. BUCKMINSTER FULLER: DISCOURSE. Edited by Joachim Krausse, Claude Lichtenstein (Editor), Zurich Museum of Design. Lars Muller Publishers, c.2001. Hardcover.

BUCKMINSTER FULLER: STARTING WITH THE UNIVERSE (WHITNEY MUSEUM OF AMERICAN ART). By Michael Hays, and Dana Miller, Yale University Press, c.2008. Paperback.

NEW VIEWS ON R. BUCKMINSTER FULLER. Edited by Hsiao-Yun Chu and Roberto G. Trujillo, Stanford University Press, c.2009. Paperback

A FULLER VIEW: BUCKMINSTER FULLER'S VISION OF HOPE AND ABUNDANCE FOR ALL. by L. Steven Sieden. Published by Divine Arts c2012

BUCKMINSTER FULLER AND ISAMU NOGUCHI: BEST OF FRIENDS. by Shoji Sadao. 5 Continents Editions, c2011.

BUCKMINSTER FULLER: Anthology for the New Millenium. Edited by Thomas T. K. Zung. St. Martin's Press, New York, c.2001. Hardcover.

Articles by R. Buckminster Fuller

1964 "The Prospect for Humanity" SATURDAY REVIEW, August 29.

1965 "The Case for a Domed City" ST. LOUIS POST DISPATCH, St. Louis, Missouri, September 26.

1966 "Vision "65 Summary Lecture" AMERICAN SCHOLAR, United Chapters of Phi Beta Kappa, Washington, D.C., Vol. 35, No. 2, Spring.

"What I Have Learned" SATURDAY REVIEW, November 12.

1967 Geosocial Revolution " SATURDAY REVIEW,September 16.

"Bucky" GRADUATE, Toronto, Canada, December. (The University of Toronto Alumni Magazine, a section entitled "Explorations" and edited by Marshall McLuhan.)

1968 "City of the Future " PLAYBOY, Vol.15, No.1 January.

"What I Am Trying To Do " SATURDAY REVIEW, March 2.

"What Quality of Environment Do We Want?" ARCHIVES OF ENVIRONMENTAL HEALTH, Vol. 16, American Medical Association, May.

"Letter to Doxiadis " MAIN CURRENTS IN MODERN THOUGHT, Vol. 25, No. 4, March/April.

1969 "The Age of the Dome " BUILD INTERNATIONAL, Vol. 2, No. 6, July/August.

"Word Meanings" and "The World Game" EKISTICS, Vol. 28, No. 167, October.

"Vertical is to Live–Horizontal is to Die "AMERICAN SCHOLAR, Vol. 39, No. 1, United Chapters of Phi Beta Kappa, Washington, D.C., Winter.

1970 "The Earthians" Critical Moment" THE NEW YORK TIMES, December 11.

1971 "A Poem by Buckminster Fuller" (Telegram sent to Senator Edmund S. Muskie of Maine), THE NEW YORK TIMES, March 21.

1972 "My New Hexa-Pent Dome: Designed for You to Live In," POPULAR SCIENCE Magazine, May.

1973 "Buckminster Fuller on Cities " THE AMERICAN WAY, AMERICAN AIR-LINES Magazine, April.

1973 "Ethics "SATURDAY REVIEW/WORLD, November 6.

1974 "Cutting the Metabilical Cord" SATURDAY REVIEW/WORLD, September 21.

"Remapping Our World," TODAY"S EDUCATION, November/December.

"Energy Through Wind Power" THE NEW YORK TIMES, January 17.

"War or Peace?" PENNSYLVANIA GAZETTE (University of Pennsylvania Alumni Magazine), April.

1975 "Time Present " HARPER'S MAGAZINE, March.

"The Meaning of Wealth" BANKER'S MAGAZINE, Vol. CCXIX, No. 1573, April.

"2025, If . . . " COEVOLUTION QUARTERLY, Spring.

"Preparing for a Small One Town World," CONGRESSIONAL RECORD--SENATE, May 22.

1976 "Preparing for a Small One Town World," THE RENAISSANCE UNIVERSAL JOURNAL, Winter.

"Buckminster Fuller" NEW DIRECTIONS, November.

"Statement of Buckminster Fuller. . ." Committee on Banking, Currency and Housing, House of Representatives Hearings on The Rebirth of the American City, Sept. 27, 28, 29, 30 & Oct. 1.

"Five Noted Thinkers Explore the Future" NATIONAL GEOGRAPHIC, Vol. 150, No. 1, July.

1977 "Mistake Mystique" EAST/WEST JOURNAL, April.

"Statement of Dr. R. Buckminster Fuller " Select Committee On Small Business, United States Senate First Session on Energy Research And Development and Small Business, May 23.

"Reach One of Those Bananas for Me" PHP MAGAZINE, Feb. 22.

"Fifty Years Ahead of My Time "SATURDAY EVENING POST, March.

"Preparing for a Small One Town World,"DHARMA, May.

1978 "Accommodating Human Unsettlement" TOWN PLANNING REVIEW, Vol. 49, No. 1, January.

"Energy Economics" JOURNAL OF EKISTICS, May.

"The Way We Live: Reflections and Projections "ARCHITECTURAL DESIGN, June.

"Future Thoughts " MOTOR WORLD, September.

"Our Physical World" CALUM (University of Calgary Alumni Magazine), Spring.

"Ever Re-Thinking the Lord"s Prayer" FINDHORN MAGAZINE, Spring. "Children Are Born True Scientists" JOURNAL OF EKISTICS, September/October.

"An Open Letter to the Architects of the World " multiple publication world-wide, November 5.

"Hang On: Here Come the "80"s" CLEVELAND PRESS, December 31.

"Old Man River" DICHOTOMY (Detroit School of Architecture) Autumn.

"Wind: The Answer?" SOLAR ENGINEERING, November.

"How Can We Have Four Billion Billionaires?" Chicago SUN-TIMES, December 2.

"How Little I Know " CHILDREN"S WORLD, Pre-Holiday Issue.

"Domes" Dad Assesses "80"s" Harrisburg (PA) PATRIOT- EWS, December 30.

1980 "An Open Letter to the Architects of the World " INSIDE/OUTSIDE (India), April/May.

1981 "Introduction to CRITICAL PATH" NEW AGE, February.

"Tensegrity" CREATIVE SCIENCE AND TECHNOLOGY, February.

"A Message of Hope to the Children of the World "NEW AGE, February (cover story).

1982 "Experiment in Individual Initiative " New Jersey BELL JOURNAL, Summer.

"Theology vs. Science: Regarding Some Familiar and Not-so-familiar Cosmic Principles "THE AMERICAN THEOSOPHIST, Fall.

Articles about R. Buckminster Fuller

1964 "The Dymaxion American," TIME Magazine (Cover Story), Vol. 83, No. 2, January 10.

1965 "Instant Slum Clearance," ESQUIRE, by June Meyer, Vol. 63, No. 4, April.

1966 "Profile," NEW YORKER, by Calvin Tomkins, January 8.

1967 "An Expo Named Buckminster Fuller," THE NEW YORK TIMES, by David Jacobs, April 23.

"The Wide World of Buckminster Fuller," PACE Magazine, November.

1969 "Entering the Age of the Unspecialist," PACE Magazine, March.

"Meet Bucky Fuller, Ambassador from Tomorrow," READER'S DIGEST, November.

"World Game," THINK Magazine (I. B.M.), November/December.

1970 "Inside Buckminster Fuller's Universe," SATURDAY REVIEW, by Harold Taylor, May 2.

"A Buckminster Fuller Survival Kit," QUEEN Magazine, London, England, May.

1971 "The View from the Year 2000," LIFE, by Barry Farell, February 26.

"Fuller is a Far-Out Guy," CHRISTIAN SCIENCE MONITOR, by Marilyn Hoffman, March 30.

"Relax—Bucky Fuller Says It's Going to be All Right," ROLLING STONE, by Hal Aigner, June 10.

"Who Will Man Spaceship Earth?" COLLEGE AND UNIVERSITY BUSINESS, September.

"Planet Earth: Buckminster Fuller's Hometown," BRITANNICA YEARBOOK OF SCIENCE AND THE FUTURE, by Robert W. Marks.

"Understanding Whole Systems," THE LAST WHOLE EARTH CATALOG, by Stewart Brand, Portola Institute/Random House.

1972 "The World of Buckminster Fuller," ARCHITECTURAL FORUM, January/February (special 40th anniversaryedition). *

Playboy Interview, PLAYBOY, February.

"Dialogue with Fuller on the Ultimately Invisible House," HOUSE & GARDEN Magazine, May.

"The World Is Flat! ` SCHOLASTIC MAGAZINE's Headline Focus Wall Map, May 1.

"Bucky Fuller's HEXA-PENT Domehouse You Can Build," POPULAR SCIENCE Magazine, June.

"Buckminster Fuller Retrospective," ARCHITECTURAL DESIGN, Edited by Michael Ben-Eli, December. *

1973 "Gladly the Dymaxion Cross I'd Bear!" by Richard Goldstein, THE VILLAGE VOICE, February 1.

"A Portrait of the Lecturer as a Whole System," SATURDAY REVIEW—Education, by Hugh Kenner, February.

"Fuller's Earth, At the Crest of the Wave," SATURDAY EVENING POST, March/April.

"Buckminster Fuller's Vision of Chicago Tomorrow," CHICAGO, March/April.

"The Synergetical World of R. Buckminster Fuller," THE QUANTITY SURVEYOR, by Sam Spencer, Part 1, September/October; Part 2, November/December.

"Fuller 2 1/2 Hours on the Universe' by Phyllis Cobbs, PATENT TRADER, March 8.

"Whole Earth Man," by Tony Lang. THE CINCINNATI ENQUIRER Magazine, November 11.

1974 "Buckminster Fuller: A Reaction," by Jim Finley. NORTHWESTERN ENGINEER, January.

"Bucky Fuller: His Ideas Mark Our Surroundings," by Thomas Hine. THE PHILADELPHIA INQUIRER, February 24.

"Bucky's Vision," by Henry Lehmann. THE MONTREAL STAR, May 25.

"Bucky Fuller in Retrospect," by Barry Brennan. EVENING OUTLOOK (Santa Monica, CA), November 9.

1975 "Here's the Man Who Wants to Make the World Work Right," by Tony Lang. EVENING NEWS (Buffalo, NY), February 15.

"Un Viaggio in Treno con Buckminster Fuller," by Gianni Pattena, DOMUS magazine (Italian), March.

"The Universe as a Scenario ' by Edmund Fuller. THE WALL STREET JOURNAL, April 8.

"Almighty Tetrahedron," by Charles Panati. NEWSWEEK, April 21.

"Ain't Nature Grand ' by O.B. Hardison, Jr. THE NEW YORK TIMES BOOK REVIEW, June 29.

"Bucky Fuller and the Final Exam," by Hugh Kenner. THE NEW YORK TIMES MAGAZINE, July 6.

"R. Buckminster Fuller," by Thierry Noyelle & Robert Wilson, METROPOLIS magazine (French), September.

"R. Buckminster Fuller, Genius," by Greg Welsh. MONDAY (Victoria, B.C., Canada), October 13.

1976 "Buckminster Fuller," DESIGN AND ENVIRONMENT, Spring.

"Geodesiculture," by Edward R. Bachtle, HORTICULTURE, Vol. LIV, No. 6, June.

"The Most Amazing Car Never Built," by Bartlett Gould, YANKEE MAGAZINE, December.

"Bringing the Universe Home," by John Poppy, est GRADUATE REVIEW, December.

1977 "Journey into Inner Space," by Stewart Dill McBride, CHRISTIAN SCIENCE MONITOR, March 9.

"Bucky 'Bored' by Carter Energy Ideas," by Susan Watters, KANSAS CITY MISSOURI TIMES, May 13.

"He'd Rather Make Sense than Dollars," by Leslie Bennetts, PHILADELPHIA BULLETIN, August 28.

"Book of the Century: Fuller's Tetrascroll" by Polly Lada-Mocarski, CRAFT HORIZONS, October.

"Inquiring for Buckminster Fuller' by Fred Kutchins, CHICAGO'S ELITE, November/December.

"Minds Instead of Muscle ' U.S. NEWS & WORLD REPORT, December 5.

"Bucky Fuller: If he were king, he would resign ," by John Corr, PHILADELPHIA INQUIRER, December 9.

1978 "Bucky Fuller is Man with Ideas' by Warren Talbot (UPI), NEW YORK CITY NEWS WORLD, January 30.

"Life Through the Mind of a Genius," by Janice Bellucci, VISTA CALIFORNIA PRESS, January 22.

"Bucky," by Carol Kahn, FAMILY HEALTH, March.

"Man Your Crystal Balls, Wise Men," by E. Patrick McQuaid, BOSTON GLOBE, April.

"Following the Trail of the Fuller Intellect," by William Marlin, CHRISTIAN SCIENCE MONITOR, May 19.

"Interview with Buckminster Fuller," by Donald Elliott, NEIGHBORHOOD, June.

"Colloquio con Buckminster Fuller," by Gianni Pettena (in Italian), MODO, June.

"Navigating Starship Earth," by John Coit, VIRGINIAN PILOT, November 15.

"Buckminster Fuller Says Mankind Faced with Final Exam," by Rollie Atkinson, FREDERICK NEWS, November 15.

"The Leonardo of Our Age," by Athena Lord, SCIENCE DIGEST, November.

"The World of Buckminster Fuller," by Ernest Ranucci, MATHEMATICS TEACHER, November.

"A Hopeful Look at the Future," by Judy Sammon, CLEVELAND PLAIN DEALER, December 1.

1979 "Transformation of a Throwaway," by Mary Earle, est GRADUATE REVIEW, January.

"The Man at the Helm of Spaceship Earth," LOS ANGELES HERALD EXAMINER, January 14.

"Our Leonardo," by Dwight Chapin, SAN FRANCISCO EXAMINER, February 22.

"Buckminster Fuller: We Have the Option of Surviving," by Don Duncan, SEATTLE TIMES, March 5.

"Buckminster Fuller," by Robert MacBride, INTERVIEW, April.

"Buckminster Fuller: An Appreciation," COMMENTARY (Singapore), April.

"Buckminster Fuller: Still Looking to the Future," by Bob German, AUSTIN AMERICAN STATESMAN, April 15.

"Making the World Work for Everyone," by Steve Parks, BALTIMORE SUN, April 16.

Associated Press Article on Tensegrity by Bruce Dallas; printed in at least 121 newspapers nationally, April.

"U.S. Inventor Takes China by Storm," by Adam Williams, SOUTH CHINA HERALD (Hong Kong), May 30.

"Buckminster Fuller in Ottawa," by Colin Alexander, SUNDAY POST OF CANADA, July 8 and 15.

"Bucky Fuller e la Prefabbricazione," by Gianni Pettena (in Italian), DOMUS, July.

"Het Overkoepelend Brein von R. Buckminster Fuller" (in Dutch), KIJK Magazine, November.

"Future Home Now Reality for Fuller," ST. PETERSBURG EVENING INDEPENDENT, November 21.

"Dr. Buckminster Fuller," by Radu Varia (in French), PARIS VOGUE, November.

"Spaceship Earth—Is It In Trouble?" by Truman Temple, EPA JOURNAL, November/December.

"You do not belong to you. You belong to the Universe." by John Love, QUEST '79 November/December.

"Futuristic Designer Has Rare Taste," TIME OF INDIA, December 14.

1980 "R. Buckminster Fuller: Genius Extraordinaire," by Julia Fung, ASIAN ARCHITECT AND BUILDER, January.

"He Makes Ideas Live," by William Miller, CLEVELAND PLAIN DEALER, February 22.

"Bucky Fuller: Um Genio em Quarto Dimensao," (in Portugese), MANCHETE (Brazil), March.

"Fuller at 85: Still Focusing on the Future' by Diana Reischel, FORT WORTH STAR TELEGRAM, April 17.

"When Fuller Talks, The World Listens' by Dick Brenneman," WEEKEND OUTLOOK, May 3-4.

"Fuller Tells Architecture Grads 'Big' Test Ahead ," by Nick Mason, BUFFALO COURIER EXPRESS, May 17.

"Help for Humanity Judged Possible," by Carol Stevens, BUFFALO COURIER EXPRESS, May 25.

"Buckminster Fuller: Entrepreneur Extraordinaire," C.A.R.E.. Digest, February.

"Buckminster Fuller: Entrepreneur Extraordinaire," C.A.R.E.. Digest, February.

"An Interview with R. Buckminster Fuller," by John Lawn, ENERGY MANAGEMENT, August/September.

"An Interview," CENTERLINE, October.

"The Man Behind the Dome," by Nance McLaren, Daytona Beach (FL) MORNING JOURNAL, October 11.

"Bucky: Thinking Makes Life Fuller," by Laura Kavesh, Orlando (FL) SENTINEL-STAR, October 23.

1981 "Progress Report on Spaceship Earth," FOCUS, January 7.

"An Encounter with Buckminster Fuller' by Frannie Noyes, Scottsdale (AZ) PROGRESS, January 10.

"Fuller Shows How to do More with Less," by Rosemary Bailey, ENGINEERING TODAY, January 13.

"Fuller's Follies," (review of CRITICAL PATH) by Hugh Kenner, SATURDAY REVIEW, February.

"Touring Spaceship Earth with Captain Bucky," by Christopher Bogan, Spokane SPOKESMAN REVIEW, February 6.

"A Terrific Package of Experience," by Christopher Bogan, Spokane SPOKESMAN REVIEW, February 9.

"Planetary Planner: In Orbit with Buckminster Fuller," by Bob Baker, Los Angeles TIMES, February 26.

"Bucky Fuller: Poet of Technology," by Linda Cruikshank, Vacaville (CA) REPORTER, March 8.

"Anne and Buckminster Fuller," by Stacey Peck, Los Angeles TIMES Home Magazine, March 15.

"The Critical Path of Bucky Fuller' by John Love, Dallas TIMES HERALD, March 22.

"Interview with Buckminster Fuller," by Wanda Urbanska, Los Angeles HERALD EXAMINER, March 25.

"Buckminster Fuller: A Non-Renewable Resource," by Joe Jenkins, NORTHWEST PRESS, March.

"Interview with Buckminster Fuller," by Jim Daley, Pennsylvania TRIANGLE (University of Pennsylvania) March-April.

"Interview with Buckminster Fuller," by Jim Daley, Pennsylvania TRIANGLE (University of Pennsylvania) March-April.

"Bucky Fuller: Courage Born of Crisis ` by Marian Christy, Boston GLOBE, April 10.

"Bucky Fuller: A Man for All Reasons," by Sheila Anne Feeney, Seattle TIMES, April 17.

"Interview with Buckminster Fuller' by Robert Anton Wilson, HIGH TIMES, May.

"Bucky at 85," The Ottawa CITIZEN, May 9.

"Buckminster Fuller: The Planet's Friendly Genius," University of Chicago MAROON, May 24.

"Fuller: Designing Genius with Plans for the Future," by Zenia Cleigh, San Diego EVENING TRIBUNE, September 18.

"The Geodesic World of Buckminster Fuller," by John Love, Detroit NEWS, September 20.

"R. Buckminster Fuller," by Beverly Creamer, Honolulu ADVERTISER, September 22.

"Fuller Discourses about Economics, Energy and God," by Ray Ydoyaga, DAILY TEXAN, September.

"Visionaries' Plans to Solve the Problem of Urban Decay," WORLD CONSTRUCTION Magazine, September.

"Bucky Fuller: Synergetic Savior," by Robert Anton Wilson, SCIENCE DIGEST, November.

"Bucky: The Leonardo of the 20th Century' by Ken McLaughlin, Watsonville (CA) REGISTER, December 1.

1982 "Buckminster Fuller on Where We're Going," by Al Tommervik, SOFTALK, January (cover story).

"Bucky Fuller's Dymaxion Transport' by G.A. Richter, Hartford (CT) COURANT, March 10.

"Buckminster Fuller: Earth's Friendly Genius at N.I.C.," by Ric Clarke,Coeur d'Alene (ID) PRESS, April 13.

"Bucky," by David Osborne, BOSTON MAGAZINE, April.

"Bucky Fuller's Global Game," by Phyllis Theroux, Washington POST, July 19.

"Coaching the World Game," by Anne Oman, Washington TIMES, July 20.

"Spaceship Fuller," by Phyllis Theroux, International HERALD TRIBUNE, July 24-25.

"The Global View of Bucky Fuller," by James McBride, Boston Globe, July 30.

"If You've Got 43 Hours ," by Virginia Inman, WALL STREET JOURNAL, July 30.

"Bucky–Folk Hero," by Jeff Cowart, Baton Rouge ADVOCATE, November 14.

1983 "The Communication Lines Are Open," by Tom Bissinger, COMMON GROUND, Winter (cover story).

*Entire issues devoted to retrospectives of Dr. Fuller's life work.

Introductions, Forewards and Contributions by R. Buckminster Fuller

Book Introductions written by R. Buckminster Fuller

MONTOSSORI AND THE SPECIAL CHILD. by R.C. Orem. G.P. Putnam's Sons, New York. c1969, Hardback.

A QUESTION OF PRIORITIES. by Edward Higbee. William Morrow and Co., Inc., New York. c1970.

CHARLES FORT—PROPHET OF THE UNEXPLAINED. by Damon Knight. Doubleday & Company, Inc., New York. c1970.

EXPANDED CINEMA. by Gene Youngblood. E.P. Dutton & Co., New York. c1970, Paperback.

A NEW LEARNING EXPERIENCE, A CASE FOR LEARNING. by Harold L. Cohen and James Filipczak. Jossey Bass, Inc., San Francisco. c1971, Hardback.

GENERATION OF NARCISSUS. by Henry Malcolm. Little, Brown, and Company, Boston. c1971, Hardback, Paperback.

CONFESSIONS OF A TRIVIALIST. by Samuel Rosenberg. Penguin Books, Inc., Baltimore. c1972, Paperback.

DESIGN FOR THE REAL WORLD, HUMAN ECOLOGY AND SOCIAL CHANGE by Victor Papenek. Pantheon House, New York. c1972, Hardback, Paperback.

SYMBOL SOURCE BOOK, AN AUTHORITATIVE GUIDE TO INTERNATIONAL GRAPHIC SYMBOLS. Edited by Henry Dreyfuss and R. Buckminster Fuller. McGraw-Hill, New York. c1972. Hardback.

CHARAS: THE IMPOSSIBLE DOME BUILDERS. by Syeus Mottel. Drake Publishers, Inc., New York. c1973, Hardback.

TM: DISCOVERING INNER ENERGY AND OVERCOMING STRESS. by Harold H. Bloomfield, M.D., Michael Peter Cain, and Dennis T. Jaffe. Delacorte Press, New York. c1975, Hardback.

NON-BEING AND SOMETHING-NESS: SELECTIONS FROM THE COMIC STRIP "INSIDE WOODY ALLEN". by Stuart Hample. Random House, New York. c1978, Paperback.

THREE MILE ISLAND: TURNING POINT. by Bill Keisling. Veritas Books, Seattle, WA. c1980, Paperback.

INDUSTRIALIZATION IN THE BUILDING INDUSTRY. by Barry Sullivan. Uniworld Industries, Los Altos, CA. c1980. Hardback.

UNIVERSITY PORTRAITS: NINE PAINTINGS. by Carolyn Plochman. Southern Illinois University Press, Carbondale, IL c1979.

CLIMATE AND ARCHITECTURE. by Jeffrey Ellis Aronin 4th edition, Van Nostrand-Reinhold, New York, NY.

DREAMRUNNER. by Jim Ballard.

Book Forewords written by R. Buckminster Fuller

DESIGNING FOR PEOPLE. by Henry Dreyfuss. Grossman Publishers, New York. c1955, 1967, Paperback.

GUIDE TO ALTERNATIVE COLLEGES AND UNIVERSITIES. by Wayne Blaze, Bill Hertzberg, Roy Krantz, and Al Lehrke. Beacon Press, Boston. c1974, Paperback.

THIS OR ELSE. by Dinshaw Dastur. Jaico Publishing House, Bombay, India. c1974.

ENERGY, EARTH AND EVERYONE. by Medard Gabel. Straight Arrow Books, San Francisco. c1975, paperback.

OUT OF THIS WORLD: AMERICAN SPACE PHOTOGRAPHY. by Paul Dickson. Dell Publishing, Inc., New York. c1977, paperback.

UNCOMMON SENSE. by Mark Davidson, J.P. Tarcher Inc., Los Angeles, CA.

ALCOHOL AS FUEL. by David Blume, published by KQED, San Francisco, CA as companion material to their radio series "Alcohol As Fuel."

Books with Contributions written by R. Buckminster Fuller

LEARNING TOMORROWS: COMMENTARIES ON THE FUTURE OF EDUCATION. Edited by Peter Wagschal. University of Massachusetts Press, Amherst, MA. c1979. Hardback.

MY HARVARD. Edited by Jeffrey Lant. Taplinger Press, New York City, NY. c1980. Hardback.

POET OF GEOMETRY PATRONS

The author would like to thank the following people for their support in helping making this book a reality.

Sue Gerst

Alan Sartirana

Buck.tv

Dylan

Ed Northrop

Elizabeth McCarthy

Anouk Nukes-Hopfer

Naomi Lou Burnett

Jean-Michel Arnoult

Tracey Daugherty and George Long

Sarah Brooks

Cari Field

Leo Natan Leicht-Gould

K.J. Hartlieb

Brent Huss

Joshua H. McManus

Nicole YIM Wing Chi

Aris Dimalanta

Patrick Gage Kelley

Pete Burness

Aaron Ruell

Lisa Pinero

Greg Tomlinson

Kirk Lancaster

Alex Odin.n

Giancarlo Canavesio

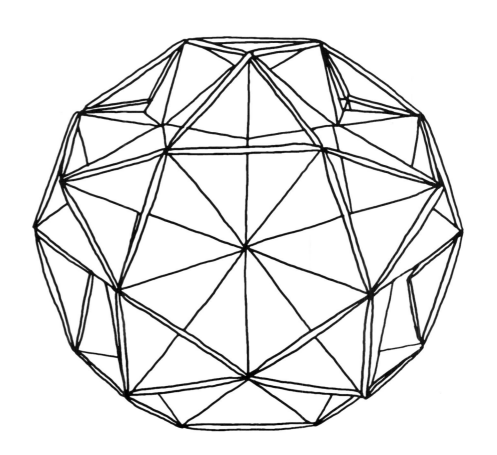

INDEX

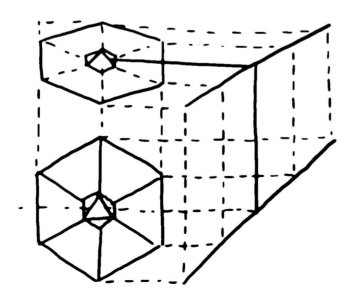